Robots

Born in Warsaw, Jasia Reichardt has lived in
England since 1946, where she is currently a
journalist, art critic, lecturer, and organizer of
exhibitions. In addition to *Robots*, she is the author
of *The Computer in Art*, among other volumes, and
the editor of *Cybernetics, Art and Ideas*. Articles by
her have appeared in *Art International*, *Zodiac*, *Art
News*, *Arts Review*, and many other publications, and
she has contributed regular columns to *Architectural
Design*, *Apollo*, *London Magazine*, *Art d'Aujourd'hui*,
Studio International, and *New Scientist*. The many
exhibitions she has organized include 'Ten Sitting
Rooms' (rooms designed by ten artists; London
1970), 'Fluorescent Chrysanthemum' (new Japanese
art, music, and films; London 1968–9), and
'Cybernetic Serendipity: The Computer and the Arts'
(London 1968, and subsequently Washington and
San Francisco). Among her many other pursuits,
Miss Reichardt is also presently a tutor at the
Architectural Association School of Architecture in
London.

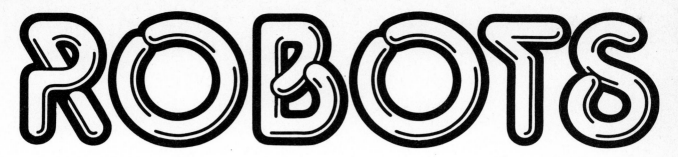

ROBOTS

Fact, Fiction, and Prediction

Jasia Reichardt

with 280 illustrations, 12 in colour

Penguin Books

'Nothing is more strange to man than his own image.'
(Alquist in *R.U.R.*, Act IV)

Drawing by Paul Flora

Penguin Books Ltd, Harmondsworth, Middlesex,
England
Penguin Books, 625 Madison Avenue, New York, New
York 10022, U.S.A.
Penguin Books Australia Ltd, Ringwood, Victoria,
Australia
Penguin Books Canada Limited, 2801 John Street,
Markham, Ontario, Canada L3R 1B4
Penguin Books (N.Z.) Ltd, 182–190 Wairau Road,
Auckland 10, New Zealand

First published in Great Britain by Thames and Hudson
Ltd 1978
First published in the United States of America in
simultaneous hardcover and paperback editions by The
Viking Press and Penguin Books 1978

Library of Congress Cataloging in Publication Data
Reichardt, Jasia.
Robots: fact, fiction, and prediction.
1. Automata.   I. Title.
TJ211.R44   1978b   629.8′92   78–8942
ISBN 0 14 00.4938 X

Printed and bound in Great Britain by Jarrold and Sons
Ltd, Norwich
Set in Monotype Times New Roman

Eduardo Paolozzi: Palucca
(statue) 1960–2, collage $7 \times 3\frac{1}{2}$ in.
(17.8×9.2 cm.)
collection: Anthony d'Offay, London ▶

The preparation of the
mechanism of an automaton
in a Paris workshop (p. 6)

I am enormously indebted to the following, whose very
generous help, hard work and encouragement made this
book possible: **Edward Ihnatowicz** of the Department of
Mechanical Engineering, University College, London, for
advice and help on technical aspects of robots; **Eduardo
Paolozzi**, who encouraged me to start the project and gave
me rare material from his collection; **Kohei Sugiura** from
Tokyo, who collected otherwise quite inaccessible Japanese
material; **Stefan Themerson**, who read the manuscript and
made invaluable comments.
I am also very grateful to the American newspapers who
published my author's query and the people who sent me
material and information in response to it: Prof. H. D. Block,
College of Engineering, Cornell University, Ithaca; Nancy
Brittingham, Independent Day School, Middlefield, Conn.;
Prof. Hubert L. Dreyfus, Dept of Philosophy, University of
California, Berkeley; Jay Dunnington, Independent Robot
Procreators, Ridgefield, Conn.; Dr David V. Forrest,
Columbia University College of Physicians and Surgeons,
New York; Andrew Ginzel, New York; Richard Gordon,
Gordon Films Inc., New York; William E. Harkins, New
York; W. Frank Heminsley, Delta, British Columbia; Prof.
William G. Lycan, Dept of Philosophy, Ohio State
University; Prof. Nils J. Nilsson, Stanford Research
Institute, Calif.; Dr Emil Oberholzer, Washington, DC; Mr
D. M. Price, Vancouver, BC; Dan Recchia, New York; Alec
Sutherland, Chicago.
Further acknowledgments and picture credits on p. 166.

Contents

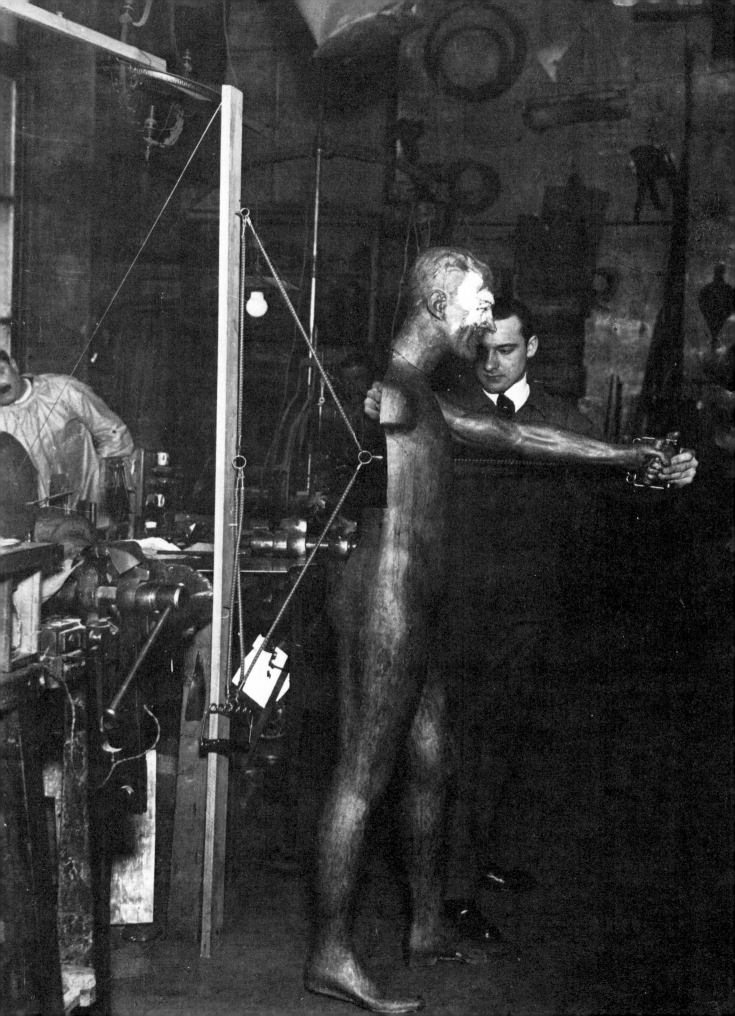

Foreword: Unfair to robots

'First, they were creeping moulds that slithered forth from the ocean onto land, and lived by devouring one another, and the more they devoured themselves, the more of them there were, and then they stood upright, supporting their globby substance by means of calcareous scaffolding and finally they built machines. From these protomachines came sentient machines, which begat intelligent machines, which in turn conceived perfect machines, for it is written that All Is Machine, from atom to Galaxy, and the machine is one and eternal, and thou shalt have no other things before thee!' This is the history of the world in epitome as related, at some unspecified time in the future, by King Armoric in a short story by Stanisław Lem.[1] According to this account, we are at present about half way through this history of mankind and machinekind, somewhere between the creation of protomachines and sentient machines, at a stage when man is still building the machines himself.

Man's preoccupation with the creation of sentient machines is the subject of this book. Usually referred to as robots, these machines have already invaded the majority of human activities: literature, film, art, sex, comics, play, war, teaching, medicine, industry, and science. Robots have become the concern of schoolboys, financiers, government ministers, trade unionists, and philosophers.

Made, at least partially, in man's image, robots have become responsible for a multitude of problems because man, having constructed the robot, has begun to distrust this new image of himself, this new impersonation of what he himself is, and what he himself does. Man's immediate reaction has been to feel an overpowering ambivalence towards this artificial creation and to turn the robot into an underprivileged sort of 'person'. And, indeed, in literature the robot has become a metaphor for a second-class citizen. Because the robot is made from metal, because it has been manufactured rather than grown, because it thinks with transistors rather than with protoplasm, it is almost always treated with condescension, even though it may demonstrate superhuman abilities, loyalty and talents.

Many robot stories are about robots which go berserk and destroy their creators. This, however, only happens when their programming contradicts the circumstances of their existence. In Karel Čapek's *R.U.R.*, for instance, when the robots rise up and kill all human beings, this is only because one of the scientists performs an irresponsible experiment to endow them with feelings, after which they can no longer abide being treated as slaves.

Isaac Asimov's 'Three Laws of Robotics' were written in 1940 in response to stories in which robots become killers. He had always pictured robots as harmless creatures doing the work for which they were designed. The robot 'was incapable of harming man, yet it was victimized by human beings who, suffering from a "Frankenstein complex" insisted on considering the poor machines to be deadly dangerous creatures.'[2] The corporation US Robot and Mechanical Men, Inc., where all the robots in Asimov's stories are produced, attempts to counter public prejudice by instituting weekly visits to the factory for anyone interested, and to encourage people to enter robotics as a life's work.

As intelligent robots become ubiquitous and taken for granted in many areas of human life, so their status as thinking beings will have to be considered. Will they have civil rights? Will they eventually be acknowledged as our equals? Will there be male and female robots? The first two questions have been pondered by philosophers for some time now. The question of gender brought forth a surprising response from children of a London school. The boys agreed that a robot is definitely 'it'; the girls said that a robot is 'he', and a few thought that some robots might be 'she', if 'she' serves a man.

The supremacy of machines over human beings is taken for granted in demonstrations of strength, accuracy, tirelessness, and reliability, but it is worth mentioning that there are other instances when people prefer machines to other human beings. In literature this occurs because machines can be extremely loyal and devoted, but in real life it may be because they do not make moral judgements.

So far the predictions about what robots will do in the home, in the factory, and in space, cannot be contradicted, although assessment as to how soon they will be able to do this has been consistently over-optimistic. Whatever the length of time required for these predictions to come true it is already obvious that man has little to fear from robots, but that he could well fear a future without them.

[1] Stanisław Lem. 'Prince Ferrix and the Princess Crystal'. *The Cyberiad*, translated from the Polish by Michael Kandel. Seabury Press, New York, 1974
[2] Isaac Asimov. 'Introduction'. *The Rest of the Robots*. Panther Books, London, 1968

A partial history

In any discussion of robots it is customary to introduce the subject with historical precedents of two sorts: artificial figures in mythology, and the development of automata. The following chronology is a selected, simplified, but representative survey of these two themes.

At the same time it demonstrates two ideas recurrent throughout this book. Firstly, that man is fascinated by the possibility of artificially created life. Secondly, that he has used all possible ingenuity to cause inanimate matter to perform the functions of living beings: whether it be human or animal, playing musical instruments, eating, or whatever else was stirring people's imagination before the time of the industrial revolution.

Such a history is only partially relevant to the advent of robots and robotics. It provides some precedents for ideas in literature, in art, films, and play, but has little to do with robots in medicine, industry, or space research. A straightforward history of medicine and technology is the only adequate background for the current developments in robotics although some ancient items from the chronology may still be of relevance, such as the automata of Hero of Alexandria. J. D. Bernal, for instance, includes mechanical toys of 600 BC — AD 500 as one of the fifteen major technical developments.

As automata at the end of the nineteenth century become increasingly toy-like. so industrial automation becomes increasingly serious. Automata and automation are as distant from each other as toy robots and robotics.

Creator or source	Automaton/Comments
God	Adam
	'And the Lord God formed man of the dust of the ground and breathed into his nostrils the breath of life; and man became a living soul.' Genesis, chapter II (This divine breath was the spiritual and intelligent soul, which distinguishes man from beasts, of which God is both the author and creator.)
	According to the Talmudic tradition, Adam is created in twelve hours, or rather twelve hours elapse between the beginning of his creation and his expulsion from Eden. 'In the first, his dust was gathered (from all parts of the world); in the second, it was kneaded into a shapeless mass (golem); in the third, his limbs were shaped; in the fourth, a soul was infused into him; in the fifth, he arose and stood on his feet . . .'.

Michelangelo. *The Creation of Adam*. Fresco in Sistine Chapel, Rome, 1508–12

Prometheus	Made the first man and woman that were ever on earth, with clay, animated by fire which he had stolen from heaven
	According to Apollodorus
Hephaestus God of all mechanical arts, also known as Vulcan, God of fire	Two female statues of pure gold which assisted him and accompanied him wherever he went
	'Living young damsels, filled with minds and wisdom'. *Iliad* (Book 18)
Hephaestus	The giant, Talus, made of brass, which guarded Crete against all intruders by heating up his body and hugging them to death
	His only vulnerable spot was his right ankle where there was a sinew of flesh and a vein of blood. The Argonauts could land on Crete only after Talus was destroyed through the intervention of Medea.

**Pygmalion
King of Cyprus**

Galatea

Pygmalion falls in love with the beautiful ivory statue which he has made, and marries her after Aphrodite brings her to life.

**Amenhotep
son of Hapu
15th century** BC

Statue of Memnon, king of Ethiopia, near Thebes in Egypt, which gave out sounds when struck by rays of the sun at dawn

This statue had the wonderful property of uttering a melodious sound every day at sunrise, reminiscent of that which is heard at the snapping of the string of a harp when it is overwound. At the setting of the sun and into the night the sound was lugubrious.
It is suggested that a divine agency was partly responsible for the sound because the mechanism was far too inadequate to sustain it.

Athanasias Kircher's representation of Memnon's statue,
from *Oedipus Aegyptiacus*, 1652 ▶

Kircher's picture of an ancient device ▶ ▶
with a bird which, when activated by the rising sun, imitated the sound and the movements of a bird

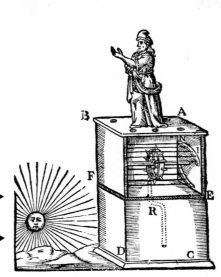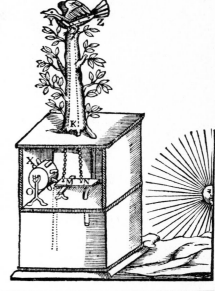

Daedalus

Moving statues worked by quicksilver which seemed to be endowed with life and walked in front of the Labyrinth, guarding it

'The animated figures stand/Adorning every public street/And seem to breathe in stone, or move their marble feet.' Pindar, *Olympic Ode, c.* 520 BC

Among other things Daedalus invented, are the wedge, the axe, the level, and a number of mechanical devices.

In Ancient Greece, statues called Daedala, made of wood and richly adorned, were carried in processions.

Talking statues have been known since 2500 BC.

Some incorporated concealed speaking trumpets through which someone hidden could address a gathering. The idea was that gods communicated through the statues which represented them.

King-shu Tse, c. 500 BC (China)

Flying magpie of wood and bamboo, and wooden horse worked by springs

Archytas of Tarentum, flourished 400–c. 397 BC

Wooden pigeon suspended from the end of a pivot which rotated by a jet of water or steam, simulating flight

Archytas is the alleged inventor of the screw and the pulley.

Ctesibius 300–270 BC

Clepsydra and water organ

Vitruvius asserts that Ctesibius '. . . discovered the natural pressure of the air and pneumatic principles . . . devised methods of raising water, automatic contrivances and amusing things of many kinds . . . black birds singing by means of waterworks and figures that drink and move, and other things that have been found to be pleasing to the eye and the ear.'

Philon of Byzantium, flourished *c.* 220–200 BC

Water automata and the repeating catapult

In his writings, fragments of which have been preserved in Arabic translation, Philon acknowledges the debt of Alexandrian pneumatics to the Egyptians who used steam and other chemical reactions in their rituals. Philon is said to have introduced the use of fire and steam, as means of producing motion, in Alexandria.

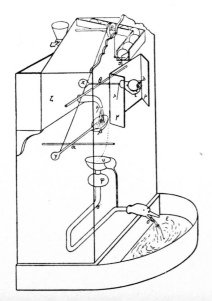

◀ **Diagram of a reconstruction of a water automaton by Philon**

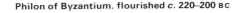

The treasury of Chhin Shih Huang Ti

A mechanical orchestra of puppets found by the first Han emperor in 206 BC

'There were twelve men cast in bronze, each 3 ft high, sitting upon a mat. Each one held either a lute, a guitar, or a mouth organ. All were dressed in flowered silks and looked like real men. Under the mat there were two bronze tubes, the upper opening of which was several feet high and protruded behind the mat. One tube was empty and in the other there was a rope as thick as a finger. If someone blew into the empty tube, and a second person pulled the rope, then all the group made music just like real musicians.'[5]

Hero of Alexandria, flourished *c*. AD 62

Treatise on Pneumatics, a famous record of ingenious applications of science which can be demonstrated by means of automata, such as his automaton theatre in which figures mounted on a box change positions before the eyes of the onlookers: singing birds, sounding trumpets, steam aeolipile (showing expansion of gas when heated and the force of reaction of its escape), animals that drink, thermoscope, siphons, and coin-operated machines

Toys were in vogue and pneumatics was used for little else. However, Hero's aeolipile toy is said to be a remote ancestor of the steam engine. He also introduced other mechanisms, which later became known as the crank, the cam-shaft, and a system of rotations of counterweights. He also demonstrated the principles of the vacuum and the incompressibility of water.

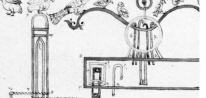

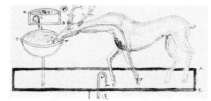

Birds made to sing and be silent alternately by flowing water. Water filling an air-tight container causes air to escape enabling birds to sing and owl to turn its back. When water is siphoned off, birds stop singing and owl turns round again.

An automaton which can be made to drink any time liquid is presented to it

A figure pouring water from a wine-skin ▶ into a full basin, without making the contents overflow

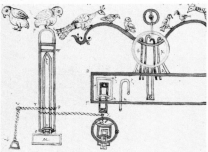

◀ On an apple being lifted, Hercules shoots a snake which then hisses.

Celsus (1st century AD), who was sceptical of magic and animals 'not really living but having appearance of life', reports that Dionysius told him that magic arts had power 'only over the uneducated, and men of corrupt morals'.

Mechanical man of jade, which could turn and move by itself. 4th century AD in China

---

Hsieh Fei

Four-wheeled sandalwood car with Buddhist statue, AD 335–45

The car was 20 ft (6.1 m.) long and more than 10 ft (3.1 m.) wide. 'It carried a golden Buddhist statue, over which nine dragons spouted water. A large wooden figure of a Taoist was made with its hands continually rubbing the front of the Buddha. There were also more than ten wooden Taoists each more than 2 ft high, all dressed in monastic robes, continually moving round the Buddha. At one point in their circuit each automatically bowed and saluted, at another each threw incense into a censer. All their actions were exactly like those of human beings. When the carriage stopped, all the movements stopped.'[5]

Huang Kun

Boats with moving figures, animals, singers, musicians and dancers, early 7th century

---

Yang Wu-Lien

A monk that stretched out its hand saying: 'Alms! Alms!', putting contributions in a satchel when they reached a certain weight, 770

Wang Chü

Wooden otter that could catch fish, 790

Prince Kaya
son of Emperor Kanmu, 794–871,
Japan

A water-spilling doll, *c*. 840

There is a tale that during a year of great drought, the prince made a doll with a big bowl and put it in his rice paddy. When the bowl filled with water, the doll would lift it up and pour it over its own face. People of Kyoto were so fascinated that they kept on filling up the bowl and watched the doll spill it. Thus the rice paddy was watered.

Han Chih-Ho

Wooden cat that could catch rats, and dancing tiger-flies, 890

'So Han took a wooden box several inches square from his pocket, and turned out from it several hundred tiger-flies, red in colour, which he said was because they had been fed on cinnabar. Then he separated them into five columns to perform a dance. When the music started they all skipped and turned in time with it, making small sounds like the buzzing of flies. When the music stopped they withdrew one after the other into their box as if they had ranks and precedences.'[5]

Scholar prince Bhoja
1018–60
India

Samarangana-sutradhara; includes comments on the construction of machines, or yantras, *c.* 1050

Among the merits of a good machine Bhoja stresses verisimilitude in representation of animals, and durability. A contemporary text relates yantras to five elements of the Universe: 'The yantras based on earth materials undertake activities like shutting doors; a water based yantra will be as lively as living organisms; a fire yantra emits flames; an air yantra moves to and fro; the element of ether serves to convey the sound generated by these yantras.'[4]

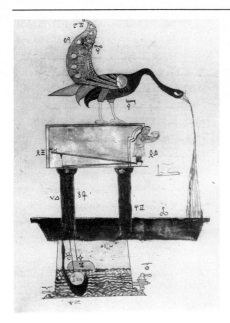
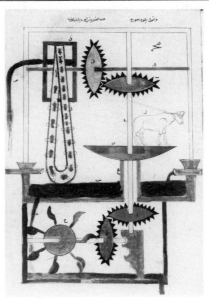

Al-Jazari at Amid on the Upper Tigris

A Book of the Knowledge of Mechanical Contrivances. A treatise describing hydraulic devices such as clepsydras and fountains, 1206

Considered by many a climax of this line of Muslim achievement. *The Peacock Fountain* is a device for washing the hands. As soon as a given quantity of water is poured into the basin, a small figure of a man emerges with a bowl of perfumed powder, soon followed by another figure with a towel.

This hydraulic pump raises water from centre right to top left of the picture. It makes use of a system of revolving shafts, wheels, and water jars hanging from a rope. The whole mechanism is powered by an ox.

Albertus Magnus
1204–72

A life-size automaton servant

There are varied accounts of what happened to the automaton. One says that when it greeted Thomas Aquinas in the street, the latter smashed it to pieces. Albertus complained that he had destroyed the work of twenty years.

According to another version, Albertus Magnus spent thirty years building an automaton of human appearance from metal, wood, glass, wax and leather, who served him. It was noted that the creature talked like a human being and opened the door to visitors. When Albertus died, Thomas Aquinas broke it to pieces believing it to be the work of the devil.

Roger Bacon
1214–94

Speaking head

The head was completed after seven years of work. It was particularly important to pay immediate attention to its words. After watching the head for three weeks, day and night, the Friars, of whom Bacon was one, handed the job over to an attendant, with the strict instructions that they were to be woken immediately the head said something. Soon after, it said: 'Time is'. It did not seem important enough to wake the Friars. A little later, the head said: 'Time was', which again was ignored. After another half-hour the head cried out: 'Time is past', and then collapsed.

Villard d'Honnecourt
13th century

Sketchbook, *c.* 1235; includes sections on mechanical devices, such as an automaton angel, and hints for the construction of human and animal figures

The angel was a statue linked by a spindle to a water-powered saw. It rotated slowly on top of a cathedral spire following the path of the sun.

Strasbourg cathedral crowing cockerel, 1352, was in action until 1789

The most famous and elaborate medieval clock. With every chime the cock appeared in the company of twelve figures, flapped its wings, raised its head and crowed three times.

Johannes Müller, known as Regiomontanus, 1436–76

Artificial eagle

It flew to greet the Emperor Maximilian on his entry into Nuremberg in 1470, while still some distance from the city, then returned to perch on top of a city gate and saluted the emperor, on his arrival, by stretching its wings and bowing.

Leonardo da Vinci, 1452–1519

Automatic lion in honour of Louis XII, *c.* 1500

As Louis entered Milan, the lion advanced towards him, stopped, opened its chest with a claw and pointed to the fleur-de-lis coat of arms of France.

Engraving from a photograph of an imitation of the Strasbourg cock

Paracelsus
1493–1541

A homunculus

A living being made by alchemy (see p. 30)

Gianello della Torre of Cremona, *c*. 1500–85

Mechanical figures of flying birds, and a girl playing the lute, *c*. 1540. Articulated soldiers who blew trumpets, beat drums and fought on the table top

The lute player walks, either in a straight line or in a circle, while plucking the strings and turning her head from side to side.
Della Torre's main task in creating these devices was to alleviate the boredom of the Emperor Charles V whom he accompanied.

Elijah of Chelm A golem, *c*. 1550

Rabbi Löw of Prague A golem, 1580

Girl playing a lute, attributed to Gianello della Torre, 1540.
Kunsthistorisches Museum, Vienna

Salomon de Caus, 1576–1626

Ornamental fountains and pleasure gardens, singing birds and imitations of the effects of nature

The apparatus used for generating bird songs and creating a theatre of natural effects follows the tradition of Hero of Alexandria.

A mechanism whereby the bird sings, but as soon as the owl appears on the rock, the ▶ song stops

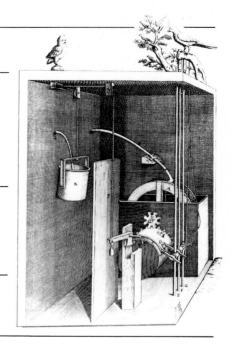

René Descartes, 1596–1650

An automaton to whom Descartes referred as 'ma fille Francine', *c*. 1640

During Descartes' voyage by sea, the inquisitive captain opened the case in which the automaton Francine was kept. On seeing her move like a living being, he threw her overboard, believing her to have been made by Satan.

Takeda theatre of automata opens 1662 in Osaka

Most automata in Japan were used in stage performances, including acrobats, fortune tellers, and calligraphers.

Christiaan Huygens
1629–95

Fountains, flying devices, coaches, music boxes

In 1680, the King of France ordered a machine showing a whole army in conflict, which was described during an address to the Royal Society. Other mechanisms mentioned included figures of artisans imitating characteristic movements of their trade, but the speaker thought it impossible to imitate human gait or voice.

Takeda Omi I

Artificial tiger which blew air from its mouth and was used as a fan, *c*. 1714

Karakuri-Kimon-Kaganigusa, 1730

An important publication about Japanese automata

Maillard

Designs for L'Académie des Sciences of chariots with cogs turning back wheel, worked by handles, 1731; and artificial swan, 1733

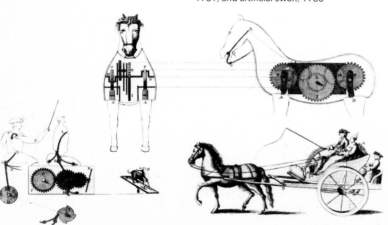

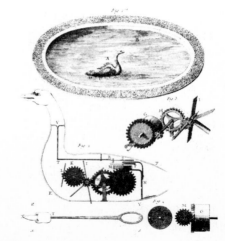

The duck – either the lost original or imitation in a ruined state.
Musée du Conservatoire National des Arts et Métiers, Paris

The flute and tabor players

Jacques de Vaucanson, 1709–82

The duck – the most famous automaton of all – was an 'artificial duck made of gilded copper who drinks, eats, quacks, splashes about on the water, and digests his food like a living duck'. It was first shown with the flute player and the tabor player in 1738.

A single wing contained more than four hundred articulated pieces and the mechanism was so complex that having been lost and fallen into disrepair, it took three and a half years to restore it so that it could perform again. A contemporary report gives the following description: 'After each of the duck's performances there was an interval of a quarter of an hour to replace the food. A singer announced the duck. As soon as the audience saw it climbing on the stage, everybody cried: "Quack, quack, quack". Greatest amazement was caused when it drank three glasses of wine.'[1]

The flute player was 5 ft 10 in. (1.8 m.) tall, on a pedestal. A current of air led through the complex mechanism causing the lips and the fingers of the player to move naturally on the flute, opening and closing holes of the instrument. It had a repertoire of twelve tunes.

People could not believe that the sounds were produced by the flute which the automaton was holding, thinking that they came from a bird organ or some other organ enclosed in the body of the figure. Most, however, were soon convinced because they could feel the breath coming from the lips of the flute player and see that the fingers determined the different notes. The spectators were allowed to see the internal mechanism in detail.

Lorenz Rosenegger

Puppet theatre of 256 figures actively representing in all detail the life of an eighteenth-century town, with every tradesman at his work. Hydraulically operated, 1752

Friedrich von Knauss, 1724–89

Automatic writing machine, 1760; talking machines and writing dolls

This writing machine was capable of writing passages of up to 107 words. It could write any text composed in advance or to dictation, when the operator would press keys on a letter keyboard.

Baron Wolfgang von Kempelen, 1734–1804

Talking machine, 1778

According to Goethe: 'The talking machine of Kempelen is not very loquacious but it pronounces certain childish words very nicely.'

Chess player, 1769

The sham automaton chess player was never actually exposed. Dressed as a Turk, the automaton sat behind a square chest 3½ ft (1.1 m.) long, 2 ft (0.6 m.) deep, and 2½ ft (0.8 m.) high. Unlocking the various doors in turn, von Kempelen would show the audience that nobody was hidden inside the chest. Then he would wind the machinery and invite a member of the audience to play. At every move made by the automaton the wheels of the machine were heard in action, the Turk moved its head and looked over every part of the chess board. When putting its opponent in check, the automaton would shake its head thrice, and twice only when the queen was threatened. It shook its head when a false move was made and replaced the opponent's piece on the square from which it had been taken; it would then make the next move itself. In general the automaton won the game. Von Kempelen admitted that there was an illusion connected with the performance of the automaton, but speculations about a minute chess master sitting in the chest have never been proved.

The fourth writing machine by von Knauss. Technisches Museum für Industrie und Gewerbe, Vienna

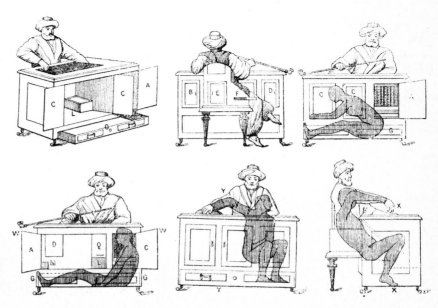

An explanatory drawing of 1882 showing how the automaton chess player might have been aided by a human, inside the chest

Pierre Jaquet-Droz
1721–90

Henri-Louis Jaquet-Droz
1752–91

The musician, Musée d'Art et d'Histoire,
Neuchâtel

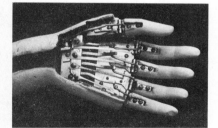

The hand of the musician

The scribe, 1770; The draughtsman, 1772; The lady musician, 1773

The scribe and the draughtsman are both boys aged about three and the musician is a girl of about sixteen. Her mechanism is divided into four parts: one, inside the keyboard instrument, operates two bellows; the other three, inside the stool on which she sits, are interconnected. 'The first, propelled by two large twin cylinders, commands a big studded brass drum made in two halves, each bearing five rows of studs corresponding to the five fingers of each hand. The two halves of this drum are separated by a set of ten steel cams. The studs on the drum act on the fingers, making them move by means of a complicated set of levers and rods which rise through the inside of the body, pass through the elbows and forearms, then through the wrists to finish up in each of the ten fingers. The steel cams compel the arm to move sideways and so bring the fingers into position above the notes that are to be played. Another mechanism operates a lever which makes the girl's chest rise and fall at regular intervals in perfect imitation of breathing. Other levers make the eyes move and animate the head and chest. Thus even when the musician pauses in her playing, she seems alive; she turns her head round and looks left and right, she casts down her eyes then looks up again, bends forward and then straightens up. At the end of a melody she does a graceful little bow.' *The Jaquet-Droz Mechanical Puppets*, Musée d'Art et d'Histoire, Neuchâtel

◀ **The scribe, Musée d'Art et d'Histoire, Neuchâtel**

The scribe's mechanism

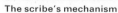

Gendaemon Wakai Tea-carrying doll, 1775

Hanzo Hosokawa

The Japanese tea doll in the eighteenth century, as described in *Karakuri Zui*, or *Illustrated Book of Mechanisms*, 1796, was constructed with cogs, springs and whale's whiskers.

Description: Doll holds a tray. If you put a tea bowl on it, the doll moves forward. If you take it off, it stops. If you replace it, the doll turns around and walks back where it came from. It moves 26 in. (66 cm.) in each direction and is 14 in. (35.6 cm.) tall. The tea doll is sometimes a male doll — since a servant could be a young boy or a girl.

Two drawings from *Karakuri Zui* showing a tea doll and its mechanism

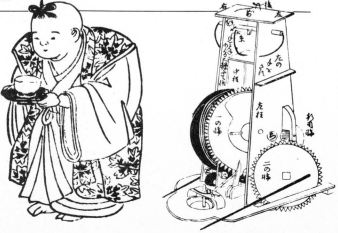

The tea-carrying doll

Following the technical description in *Karakuri Zui*, a tea doll was satisfactorily reconstructed recently by a group of students in Tokyo.

Karakuri Zui explained mechanics to the public at large.

The introduction to the book says that the creation of mechanisms must be based on observation of natural phenomena and conducted in the same spirit. Seeing the fish move its tail in the water — think of the rudder of a boat; when the fish moves its fins — think of the oars.

Igashichi Iizuka, 1762–1836   Sake-carrying doll, *c.* 1790 (see p. 112)

Tippoo's Tiger, *c.* 1792

Made after the son of Sultan Tipu's enemy was savaged by his favourite creature, the tiger, on Sangor Island in 1792. While the movement is limited to the victim's forearm, the tiger's attack is accompanied by fearful growling.

Carved wooden tiger, containing miniature organ, made for Sultan Tipu, and captured by the British during the fall of Seringapatan in 1799. Victoria and Albert Museum, London

---

Johann Nepomuk Maelzel 1776–1855

The metronome (perfected and patented); also Pan-Harmonikon, an apparatus imitating a whole orchestra with forty-two separate automatons. A trumpeter, 1808, wearing French or Austrian uniform, which played march tunes of each country, according to its attire.

Johann-Gottfried and Friedrich Kaufmann

A trumpeter, 1810

Described by Carl Maria von Weber as 'automatic and properly blowing life-size trumpeter'.

Les Maillardet: Henri, Jean-David, Julien-Auguste, Jacques-Rodolphe active late 18th and early 19th century

Draughtsman–writer, figure of a boy kneeling on one knee with a pencil in his hand. Wrote in English and French and drew landscapes. Before 1812

Ended up as a girl, kneeling on both knees, at the Franklin Institute in Philadelphia.

The Philadelphia doll

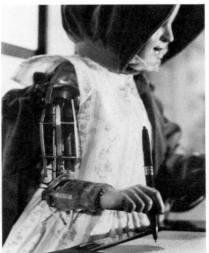

The doll's arm mechanism

A drawing by the Philadelphia doll

A mechanical magician which answered questions, which were placed in a drawer from which an answer was later retrieved

Questions and answers:
'What is the most noble reward of knowledge?' — 'To enlighten ignorance'.
'How should we think of morality?' — 'As the hygiene of the soul'.

Humming bird in a box

When the lid was opened the bird flew out, fluttered its wings, opened its beak and warbled four different songs. It then returned to the box and the lid closed.

Joseph Faber, *c*. 1800–50

Speaking automaton, Euphonia, 1830

Euphonia was exhibited at the Egyptian Hall in Piccadilly in 1846. The invention of Professor Faber of Vienna, it took twenty-five years to construct. Euphonia was a bearded Turk, which is why it was sometimes called Euphonis. The automaton produced sounds similar to the human voice: it started by reciting the letters of the alphabet and then said 'How do you do, ladies and gentlemen?' It asked and answered questions, whispered, laughed and sang. The mechanism could be inspected and it contained a double bellows, keys, and levers. The Turk had a mouth which moved, a tongue, and an indiarubber palate. Since it was built in Austria by a German speaker, the automaton spoke English with a German accent. The mechanism of Euphonia was inspected on many occasions to everyone's satisfaction that the sounds were produced by the machine and that no ventriloquism was involved.

Robert Houdin, 1805–71

Writing doll which signed Houdin's name, 1840. Pastry salesman, acrobat, tight-rope dancers, marksman, and a trapeze artist

◄ **The marksman**

Benkichi Oono, 1801–81

A doll expressing feelings

Oono was ordered to make a doll by a feudal lord. When it was finished, it proceeded on its knees towards the lord, carrying a tea bowl. Pleased with what he saw, the lord hit it on the head with a fan. At this point the doll stood up, glared at the feudal lord, and made a gesture of drawing a sword from the scabbard. Benkichi was never forgiven although the doll was only a toy. The feudal rulers suspected the doll makers on a number of occasions of undermining their authority.

Alva Thomas Edison 1847–1931

Talking dolls, 1891

Initially made to advertise the phonograph, they became the rage.

Mechanism of a wind-up walking doll, *c*. 1870

George Moore

The steam man was powered by a 0.5 horsepower gas-fired boiler. It reached a walking speed of 9 mph (14 kph) and used its cigar as a steam vent. 1893

Described as a walking locomotive

Skipping doll, and walking elephant worked by a parasol-driven fly-wheel, 1890

Among the books consulted in constructing this chart are:

1. Alfred Chapuis and Edmond Droz. *Automata – a Historical and Technological Study*. Editions du Griffon, Neuchâtel, 1958
2. John Cohen. *Human Robots in Myth and Science*. George Allen & Unwin, London, 1966
3. Mary Hillier. *Automata and Mechanical Toys*. Jupiter Books, London, 1976
4. Gustav Metzger. 'Automata in History: Part 2'. *Studio International*, vol. 178, no. 915, October 1969
5. Joseph Needham. *Science and Civilisation in China*, vol. 4, part II. Cambridge University Press, 1965
6. René Simmen. *Mens & Machine*. Uitgeverij van Lindonk, Amsterdam, 1968
7. Shoji Tachikawa. *Karakuri*. Hosei University Press, Tokyo, 1969

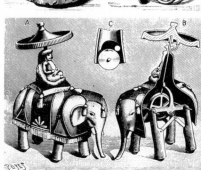

The Stanford Research Institute ▶ robot vehicle Shakey was the first complete robot system (see also p. 156).

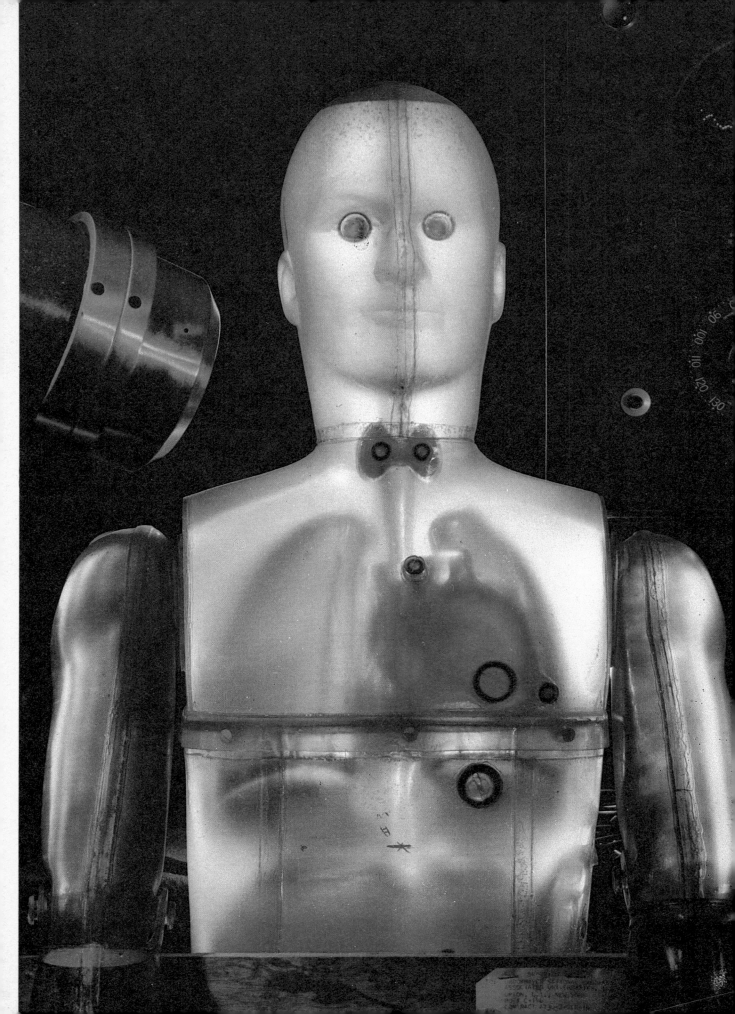

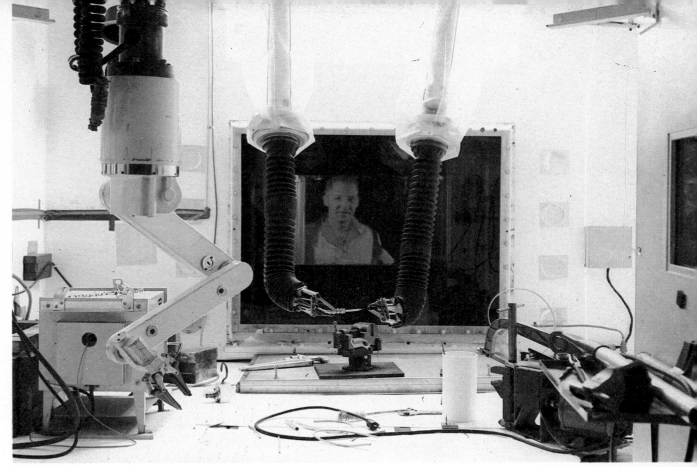

Plastic robot with artificial organs, from Brookhaven National Laboratory, for monitoring radiation doses. It has comparable radiation absorption to a human being.

Unimate industrial robot: highly competent at welding, spraying and machining, it can make 2000 welds per hour, and work sixteen hours a day.

Master/slave manipulators used for work with radioactive substances. The operator is behind a protective tank containing zinc bromide solution.

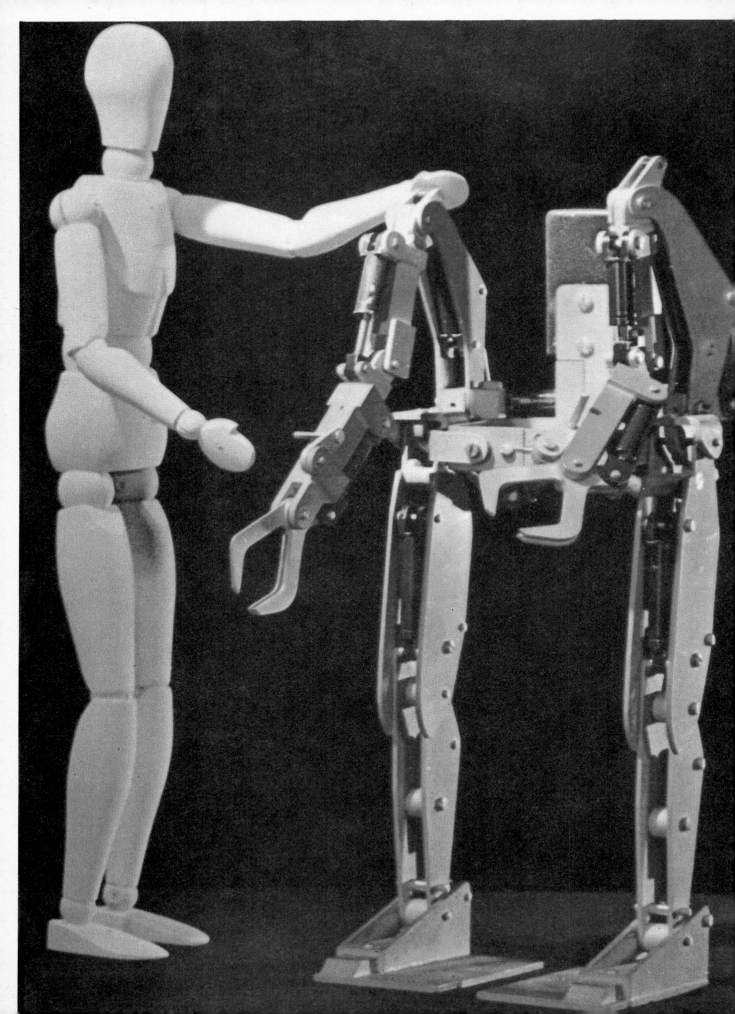

Hardiman, an exoskeleton for amplifying human force. A man ◀ inside the mechanical frame can lift weights greater than 992 lbs (450 kg.)

The JPL/Ames anthropomorphic manipulator can be controlled by the master arm worn by operator, manual control console, or computer. ▲

Voice-commanded wheelchair with an arm to aid quadriplegics: an application of teleoperator-robot research to rehabilitation engineering ▼

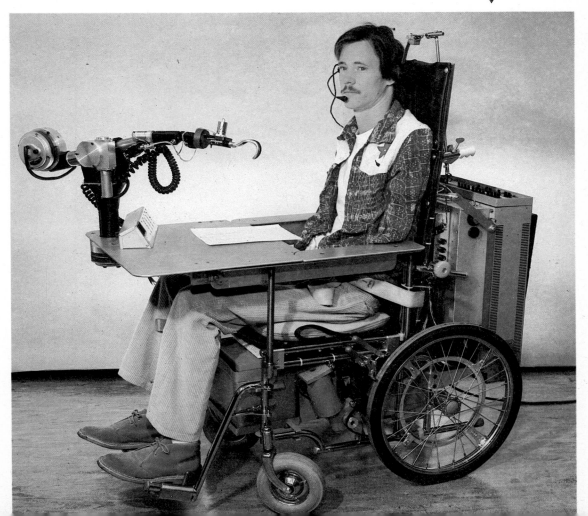

◄ Performing robot at the Fujimura
Robot Pavilion in Tokyo

Toy robots from the collection of Jean
Philippe Lenclos, Paris
▲

Children's performance of a robot play
They Came from the Sun
▼

▲

Robot stars C3PO and
R2-D2 from *Star Wars,* the
biggest commercial success
in the history of the cinema.
Director George Lucas
called his protagonists
'strange creatures in a
strange land' but audiences
do not find them in the least
strange even though See
Threepio speaks squeaky
English and Artoo Deetoo
speaks Electronic.

Robot from the long-
running BBC television
weekly serial *Doctor Who.* A
science fiction programme
for children aged eleven to
fourteen, it is also very
popular with many adults
and has been presented
since 1963.

▶

Human! Don't be fooled!

The world of automata and robots contains an area of illusion and fraud which presents many traps for the innocent. The Vaucanson duck might have seemed as miraculous in its performance of eating and digesting as the chess-playing Turk at chess, but whereas the mechanism of the former was available for inspection, the von Kempelen chess player could only be examined under certain conditions, and that is by being demonstrated by von Kempelen himself.

In recent years robots and illusions have also become confused. A female robot demonstrating equipment at an Electronics and Automation exhibition in London in 1968 turned out to be a living girl connected by wires to what was ostensibly a computer console. More recently still, an advertisement for a robot chef referred to a robot capable of cooking eggs. There are many machines called robots which are as autonomous, electronic, or cybernetic, as a water tap once it has been turned on.

Comments on commercial exploitation, man's desire for magic and instant solutions to problems, as well as his credulity are to be found in literature. Robert Escarpit in *The Novel Computer*[1] satirizes just this subject. The book concerns Literatron, a literary computer which initially does not exist although pictures of it are reproduced in magazines. The fact that what is actually reproduced is a washing machine and not the computer does not seem to have been noticed by more than one person. The suffix 'tron', *vide* positron, cyclotron, betatron, is established to be extremely useful: 'One well-placed tron, and you'll scoop up millions in grants. . . . Say you invent a robot to work in the kitchen – a really revolutionary, fantastic, faultless discovery. No one will offer you anything to develop it. But present it as a magitron – from *mageiros*, you know, the Greek for kitchen – and you'll find yourself showered with gold.' Literatron was put to work on the production of best-selling novels such as *The Virgin Typist*, and some blank verse called *What Boots It* which was rendered in several versions including one for children. There were other trons, including an ecumenotron which was working out plans for a syncretotron, 'designed to unify all systems of theology by mechanical means'. One of the proposed systems which had not yet been called by any name was an automatic confessional: 'There's a dial with a pointer you turn to the various sins. You put half a dollar in the slot and get a ticket of absolution, which you turn in at the sacristy against so many candles or prayers or what-have-you.' Nineteen centuries separate Escarpit's coin-operated confessional and Hero of Alexandria's first coin-in-the-slot machine. An impartial observer from outer space might well ask: 'Is it progress?'

However, Escarpit's trons were not as dangerous as one of the most desirable robots of all time, whose pronouncements were never true and yet whose every word was always believed. This robot was RB 34, known as Herbie, which appears in Asimov's story 'Liar!'[2] US Robot and Mechanical Men, Inc. quite unintentionally produced a robot capable of reading minds. Furthermore, Herbie did not want to read science but got interested in fiction, especially romantic novels, which provided him with plenty of scope for the commentaries which he generously expounded to anyone who wanted advice, counsel, or who would simply just listen to him. He told people exactly what he supposed, or deduced, they would like to hear and in the end was responsible for creating the worst havoc US Robots had ever experienced.

In robot lore, truth as a concept may not seem the most relevant or vital criterion, but fraud in automation is worse than human deception because its association with science makes it seem impervious to corruption.

Drawing of Rabot by Al Lorenz[3]

Rabot (RAY-bott). (From the Walloon *rabbett*, young of the cony; also Sanskrit *rabhas*, violence, force)

Back in twenty-first-century America, a species of rabbit that crossbred with copying machines. In the worldwide famine that followed the unwinning of the war against the Rabots, the last ream of carbon paper was consumed and history returned to its original oral tradition.

[1] Robert Escarpit. *The Novel Computer*. Secker & Warburg, London, 1966 (first published by Flammarion, Paris, 1964)
[2] Isaac Asimov. 'Liar!'. *I, Robot*. Panther Books, London, 1968 (first published in *Astounding Science Fiction*, 1941)

[3] Eve Merriam. *Ab to Zogg – A Lexicon for Science-Fiction and Fantasy Readers*. Atheneum, New York, 1977

Human reactions to imitation humans, or Masahiro Mori's Uncanny Valley

Mori's[1] imaginative interpretation of the relationship between man and robot is expressed in two graphs.

The thesis is that the closer the robot resembles a human being, the more affection or feeling of familiarity it can engender. Contrary to what one might expect, however, the imitation of human exteriors may lead to unexpected effects and unpleasant surprises.

An industrial robot, with neither face, nor legs, and only arms and hands that move and manipulate objects, is the direct outcome of functional design. It does not look human in any way, and although it has hands, or hand-like extremities, it would rate low on the scale of familiarity on the graph, and would therefore be unlikely to arouse affection in a human. On the other hand, with toys the shape is more important than function, and so a toy robot will have a face and limbs because it is designed to arouse affection in children.

There is a school of thought that robots should be made to look like, and function entirely like, human beings; a good example of this principle is the latest prostheses which resemble human limbs as closely as possible. Artificial arms, which used to consist of iron rods held together by nuts and bolts, have become soft, fleshy and skin-coloured, and at first glance look almost like real ones. The design of artificial hands has progressed so far that one can even see veins and tendons; the fingers have nails and can produce fingerprints. The colour of the skin is slightly more pink than real skin, or as real skin might be after a hot bath. The difference in appearance between a real human hand and an artificial one is now no greater than that between real teeth and artificial teeth. Ironically, artificial hands look so convincingly real that people get an uncanny feeling when they find out that they are artificial. 'If you shake an artificial hand you may not be able to help jumping up with a scream, having received a horrible, cold, spongy grasp', writes Mori. Once you know that the arm is artificial you feel no sense of familiarity with it whatever. On the graph, Mori would place such an artificial arm near the bottom of the uncanny valley.

A performing Bunraku puppet looks convincingly real from the seats in the audience, especially during those moments when the subtle movements of the eyes and hands create an illusion of life. It would therefore be given quite a high place on the graph. Of course, everybody knows that the Bunraku puppet is only a doll and so no one's expectations can be disappointed because the spectators actively participate in the illusion.

Motion, or lack of it, has a very definite effect on where things are placed on the graph. An industrial robot in motion comes closer to a human hand when it performs the movements of a hand than when it is switched off and is just a static piece of machinery smelling of oil. The difference between the industrial robot and the artificial hand which imitates a real one is that we recognize the robot for what it is and we are tricked by the artificial hand. When an artificial arm of the most complex type, which can revolve and which is surmounted by a hand with fingers capable of opening and closing, suddenly stops functioning, this creates a feeling of unease in those who witness it. Mori compares such an experience with that of an engineer who goes into his workshop in the middle of the night and finds that the manikin dolls he left behind have suddenly started to move.

Movement where we anticipate stillness and stillness where we expect movement is upsetting. Even the speed of movement can change the effect and drastically counteract our expectations. One of the robots at the 1970 Osaka World's Fair incorporated twenty-nine facial muscles so that a range of expressions could be programmed, the idea being to make the robot smile. According to the engineer who made it, a smile is a sequence of facial distortions performed at a certain speed. If the speed is decreased by half, the robot's smile, far from being charming, would become a frightening and uncanny grin. This is another example of the strange effect caused by something unexpected. If anything at all goes wrong, either with an artificial hand, or a doll, or a smiling robot, those witnessing the event shudder, and the particular devices immediately drop into the uncanny valley.

We assume that the harder an accelerator is pressed, the faster the car will travel, but even that is not always the case, and the shortest route is not always the quickest. The most obvious system of cause and effect is far too simplistic to influence the design of such extremely subtle things as robots. Mori suggests that the design of robots should be reconsidered and that they should not depend too much on external similarity to human beings just because they have to perform human tasks. He proposes a position for robots on the graph which would be somewhere near the top of the hill on the left, denoting a reasonable degree of familiarity which can be achieved and maintained. He gives the example of spectacles, of which the design makes it obvious that here is a man-made addition to a face (often capable of making it more attractive), and suggests that instead of a pathetic-looking humanoid artificial hand, there should be a smart, obviously false hand, one that would be soft and gentle with beautifully shaped curves.

Other elements shown on the second graph are there to illustrate the range of degrees of familiarity and their effect: the unexpected touch of a cold, spongy, artificial hand is not at the very bottom of the uncanny valley; for Mori the most terrifying and the deepest secret of the uncanny valley is a moving corpse, which would be even more frightening than the most realistic artificial hand going awry.

What changes the status of an object or a person, and reverses our feelings towards them, can be just one single variable. Mori's analysis in relation to robots shows how to guard against disappointing consequences.

[1] Professor Mori heads the Robotics Department of the Tokyo Institute of Technology.

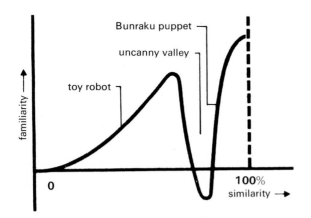

Graphs of the relationship between machine and human being

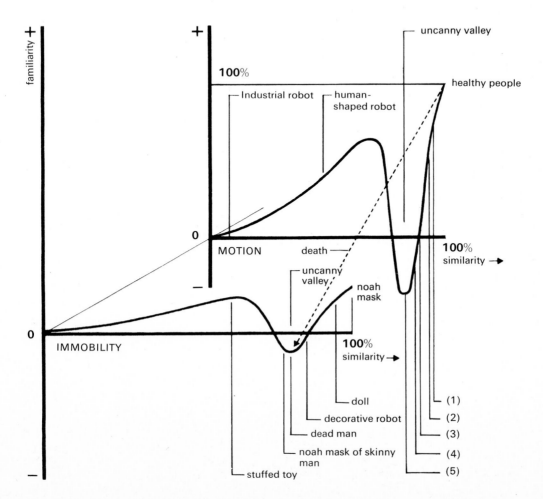

(1) unhealthy people
(2) Bunraku puppet
(3) handicapped people
(4) artificial electric hand
(5) moving dead man

Definitions

The definitions of those artifacts, in fact and fiction, which imitate real people, differ only in shades of meaning. They all have one thing in common. Their purpose is to benefit man, by serving, protecting and entertaining him. Man seeks those qualities in the things-he-makes-in-his-own-image which he cannot usually get from his fellow humans.

Android

From the Greek 'andros' meaning of man, and 'eidos' meaning form. The name given to any machine constructed to imitate the appearance and actions of man. The Vaucanson flute player of 1738 (see p. 13), for instance, was referred to as an android. The contemporary meaning of the term is somewhat different. An android is a robot with human form whom one cannot tell apart from a human being. *The Visual Encyclopedia of Science Fiction*[1] says: 'Androids may be defined as "robots made of flesh". While they can be programmed to accept orders in the same way as robots, their bodies are chemically or biologically based and are grown rather than built.' In the television film serial *Lost in Space*, the beautiful female android Varda comes out of a vending machine. Hadaly, the android in *L'Eve future* by Villiers de L'Isle Adam, is the result of a collaboration between an engineer and a medium who infuses a spirit into the inanimate hardware.

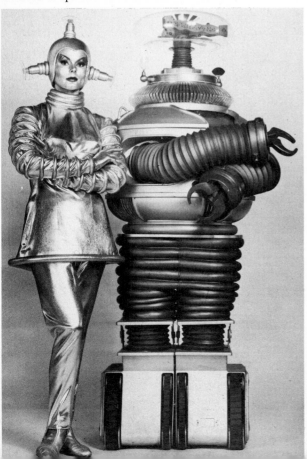

Automaton

From the Greek 'automatos', acting of itself. Automata, often highly decorative, are mechanical artifacts which tend to imitate things from real life. Based on clock mechanisms they continue to move and perform, once they have been set in motion. *Encyclopaedia Britannica* omits robots from its definition of automata because robots are defined as being functional, which automata are not. The *Concise Oxford Dictionary*, 1976, says: 'Piece of mechanism with concealed motive power; organism, esp. person, whose actions are involuntary or without active intelligence.' In metaphysics, there is the 'theory that animal and human organisms are automata, that is to say, are machines governed by the laws of physics and mechanics. Automatism, as propounded by Descartes, considered the lower animals to be pure automata, and man a machine controlled by a rational soul. Pure automatism for man as well as animals is advocated by La Mettrie' (1748).[2]

Cyborg

This is another term for Bionic Man – a being who is part-machine and part-flesh, a synthesis of nature and technology. Originally, it comes about as a result of a tragic accident which causes the person to be rebuilt with spare parts. The best-known cyborg is The Six Million Dollar Man, based on the novel *Cyborg* by Martin Caidin. An astronaut saved from a crash undergoes mechanical repairs and emerges, still in human form, but with a bionic arm, bionic legs and telescopic vision. This serial was so successful that it was followed by another: The Bionic Woman.

Android, Varda, with robot, from *Lost in Space*, 20th Century–Fox television serial, 1966

Cybermen from the BBC television serial *Doctor Who*, 1968

A robot in which a human brain has been implanted is also a cyborg. The cybermen of the BBC television serial *Doctor Who* also belong here. They were once men whose bodies, as they became old and diseased, were replaced limb by limb with mechanical parts. Finally, 'even the human circulation and nervous system were recreated, and brains replaced by computers'. The main problem with cyborgs is the problem of identity. In *Who?*,[3] a US nuclear scientist is rebuilt by the Russians, but nobody can be quite sure whether he is who he says he is, or whether he is someone else programmed to pretend to be him, because the part of him which is missing is his face, and his head is a polished metal ovoid which is featureless except for a grille in place of the mouth. Another example is given by Stanisław Lem in his story 'Are you there, Mr Jones?', 1969, in which a racing driver is gradually rebuilt after a series of crashes. He cannot meet the bills for his reconstruction, but when it is established that every organic part has been replaced, he cannot be sued because a robot cannot be brought to trial. Until all his parts were replaced he was a cyborg; when there was nothing left of his original self, he became a robot.

Golem

The term is used in the Bible and in Talmudic literature to refer to an embryonic or an incomplete substance. In the creation of Adam, at the third of the seven stages, before he finally came to life when God breathed into his nostrils, his state was described by the rabbis as that of a golem, i.e. a shapeless, unformed, substance.

There have been, since the Middle Ages, many stories about wise men who made human effigies from the dust of the earth and then brought them to life with a charm or a shem, or a tetragrammaton of the four letters of God's ineffable name written on a piece of paper and placed in the golem's mouth. The earliest named person to be credited with the making of a golem is Elijah of Chelm in the middle of the sixteenth century. His grandson was so convinced that this legend was true that he discussed the theologically legal question of whether a golem could be one of ten men needed for the performance of a religious service. There is also a legend about Ben Sira who studied the book of creation in order to make a golem, which he succeeded in doing in three years. He brought the golem to life by writing 'emeth' (truth) on his forehead. But on learning that the same word had been placed on Adam's forehead to bring him to life, the golem begged him to take it away because he did not want to go astray like Adam. One letter of the word was removed so that it read 'meth' (dead) and the golem turned to

Scene from *Der Golem*, 1920

dust. Each century had its golems; the last one was said to have been made in the early nineteenth century in Grodno, Russia.

Meyrink's novel of 1915, *The Golem*, is based on dreamlike sequences and curious events such as disappearances, plagues, and murders, all of which are attributed to the power of the golem which lurks and waits around the ghetto in Prague. The golem is treated like a legend until something happens and it turns into actuality once more. Then, again, it is just remembered as a spirit craving materialization. Meyrink writes: 'With the help of an ancient formula, a rabbi is said to have put together an automatic man and used it to help ring the bells in the Synagogue and for all kinds of other menial work. But he hadn't made it into a proper man; it was more like a kind of animated vegetable, really. What life it had, too, so the story runs, only derived from a magic prescription placed behind his teeth each day, that drew down to itself what was known as "the free sidereal strength of the universe". And as, one evening, before evening prayers, the rabbi forgot to take the prescription out of the Golem's mouth, the figure fell into frenzy, and went raging through the streets like a roaring lion, seeking whom it might devour. At last the rabbi was able to secure it, and he then destroyed the formula. The figure fell to pieces. The only record left of it was the miniature clay figure that was shown to the people within the old Synagogue.'[4]

[1] Brian Ash, ed. *The Visual Encyclopedia of Science Fiction*. Pan Books, London and Sydney, 1977
[2] Ledger Wood in *The Dictionary of Philosophy*. Philosophical Library, New York, 1942
[3] Algis Budrys. *Who?*. Ballantine Books, New York and Toronto, 1975
[4] Gustav Meyrink. *The Golem*. Victor Gollancz, London, 1928

Homunculus

A homunculus is made with the aid of alchemy. Paracelsus (1493–1541), in a lecture delivered to the faculty at the University of Basle, gave the following recipe: 'Let the semen putrefy by itself in a cucurbite with the highest putrefaction of the *venter equinus*, for forty days, or until it begins to live, move, and be agitated, which can easily be seen. After this time it will be in some degree like a human being, but, nevertheless, transparent and without body. If now, after this, it be every day nourished and fed cautiously and prudently with the arcanum of human blood, and kept for forty weeks in the perpetual and equal heat of a *venter equinus*, it becomes thenceforth a true and living infant, having all the members of a child that is born from a woman, but much smaller. This we call a homunculus; and it should be afterwards educated with the greatest care and zeal, until it grows and begins to display intelligence.'

In Goethe's *Faust*, part 2, Mephistopheles watches the making of a homunculus by Faust's assistant Wagner. In a medieval laboratory a translucent form of the homunculus appears in a vial which before was merely full of froth. Homunculus speaks of the limitations of man-made creations by comparison with those of nature, and goes on to say that while he lives he must work and start on his tasks without delay.

WAGNER: . . . My mannikin! What can the world ask more?
The mystery is brought to light of day.

N. Hartsoeker. Drawing of a human spermatozoon containing a homunculus, or what he imagined a spermatozoon would look like if it were possible to see it clearly. 'Essay de Diptrique', 1694

Now comes the whisper we are waiting for:
He forms his speech, has clear-cut words to say.

HOMUNCULUS: . . . That is the way that things are apt to take:
The cosmos scarce will compass Nature's kind,
But man's creations need to be confined.
. . . For while I live I must be up and doing.
I'll brace myself to work, without delay, . . .

'The theme of creation of life by scientific synthesis is seen in the birth of Homunculus, a brain-spun creature longing for organic life and higher development: the glass bearing his aspiration, its light and music, is at last shattered at the throne of love, . . . years before Darwin, he (Goethe) showed clearly a similar scientific outlook:

Then, following eternal norms,
You move through multitudinous forms
To reach at last the state of man.'[5]

In Laurence Sterne's *Tristram Shandy*, Tristram's state before the emergence into the world is described as that of a homunculus. '. . . for nine long, long months together – I tremble to think what a foundation had been laid for a thousand weaknesses both of body and mind, which no skill of the physician or the philosopher could ever afterwards have set thoroughly to rights.'[6]

Homunculus usually refers to a being organically but artificially grown and is always used in relation to an embryo or a newly emergent life.

Robot

A machine which wholly, or in part, imitates man – sometimes his appearance, sometimes his actions, and sometimes both. (See also chapter 'To work! To work!', p. 138.)

At one extreme, in fiction, a robot can replace man and even better him. Although robots are not supposed to have feelings they often manifest them and insist that they are human, or at least that they are not machines. In real life, a robot does a part of the job of some parts of a human anatomy. The computer can be said to represent a man's head; the industrial robot his arms; and the walking machines his legs.

The term was coined by Karel Čapek in 1917 and

[5] Philip Wayne, translator of *Faust*. Penguin Books, London, 1959
[6] Laurence Sterne. *The Life and Opinions of Tristram Shandy, Gentleman*. Basil Blackwell, Oxford, 1926–7, as part of complete works

was first used in his short story 'Opilec', which means drunkard. The word 'robot' comes from the Czech 'robota' which (unlike the same word in Yugoslav and Polish which merely denotes work) means obligatory work or servitude. It derives from the period of feudalism and the labour which peasants owed to their masters for their accommodation and a patch of land. It next appears in Čapek's famous play *R.U.R.* or *Rossum's Universal Robots* (see p. 36), first published in Czechoslovakia in 1920, first performed at the National Theatre in Prague on 25 January 1921, and first published in English translation in 1923. In *R.U.R.*, a robot is defined as an artificial humanoid machine turned out in a factory in great numbers and sold as cheap labour.

Čapek's play has given an overall name to an entire range of objects: mechanical and electronic humanoids, androids, and manikins whose original purpose is to serve man but whose ultimate purpose might be something quite different.

**Hans Küchler.
Robot suicide.
From *Mechunculi robotenses*.
René Simmen, Zurich, 1967**

Teraphim

Hebrew word found only in the plural, of uncertain etymology, referring to household gods of the ancient Hebrews. They were humanoid in form and were made in great numbers, a place being kept for them in every household since they were not inconsistent with the worship of God. They were consulted as oracles by the Jews. In 620 BC they were put in the class of idols and eventually banned by the prophets. Some incorporated mummified heads but it is not even certain what they looked like.

In this representation by Athanasias Kircher in his *Oedipus Aegyptiacus*, 1652, they are described as follows:
A. Water-clock teraphim
B. Portable teraphim
C. Teraphim approximating to the image of Horus of the Egyptians
G. Head of the first-born on the wall, and
F. Golden plate under the tongue
Teraphim, like the golem, represent man's effort to create an image in his own likeness to help him.

Teraphim Hebraeorum from Athanasias Kircher's *Oedipus Aegyptiacus*, 1652

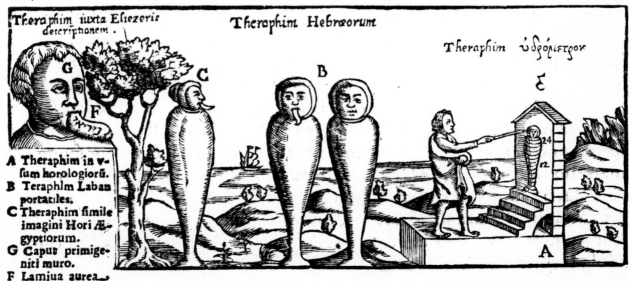

In search of perfection

Man's search for an ideal woman has been immortalized in numerous stories, many of which are remarkably similar. There are three main types of artificial women in fiction: a romantic and dreamlike woman endowed with spirit, wit and intelligence; a practical household companion; and a passive doll whose great virtue, by saying practically nothing, is to become a flattering mirror for the man who falls in love with her. All three possess beauty, elegance and charm.

The most remarkable artificial woman is the invention of Villiers de l'Isle Adam who wrote his 'L'Eve nouvelle' in 1879, which has subsequently become famous as *L'Eve future*.[1] The story is that of a young Englishman, Lord Ewald, who falls in love with a singer and comedienne, Alicia Clary, whose beautiful exterior is unmatched by her vulgar, common, and silly soul. A solution is offered by the electrician, Thomas Alva Edison, who wants to create for his friend an *Andréide* which will be the exact double of Alicia physically, but not spiritually. The book deals at great length with the process by which Hadaly[2] is made. Edison is responsible for the mechanism which is animated by electricity and which is imbued in a mysterious way with soul or spirit by a woman of supernatural powers. Edison refers to the process as being less a transformation of inanimate matter than an experiment with photo-sculpture. On meeting Hadaly, Ewald cannot believe that he is not talking to Alicia. Gradually he begins not only to love but also to accept the extraordinary qualities of Hadaly as a person, her gift of sensibility and poetic response to life, and soon decides to take her back with him to England. When the casket in which she is 'asleep', waiting to be reanimated on her arrival, is burnt with the ship, we learn from a newspaper that Ewald had offered a fantastic sum of money to anyone who would have saved an elegant box from the flames. A day later Edison receives a cable from Ewald: 'I shall not grieve except for this shadow.'

According to Villiers' first biographer,[3] there are speculations that Villiers was present at a dinner during which someone told a story of a young Englishman who committed suicide and was found dead in bed, spattered with blood, next to a wax effigy of a girl from London renowned for her beauty. A young American engineer who was present said quietly that he regretted not having known about this tragic love story before it was too late as he could have prevented the outcome. When everyone asked him how it could have been done, he said 'By imbuing the doll with life, soul, movement, and love'. The present company was sceptical of such miracles and everyone laughed, except Villiers. 'You can laugh', said the American gravely, 'but my master, Edison, will soon teach you that electricity is as powerful as God.' It could be that from this nocturnal conversation was born one of the most original French novels of the end of the nineteenth century. However, this report may or may not be accurate. Elsewhere it has been claimed that Lord Ewald is Villiers himself and that the story is autobiographical.

During the fifteen years before the story was written, Thomas Alva Edison (after whom the electrician of the book is named), invented the phonograph and the electric light bulb. He himself became the object of fantasy by being called the Magician of the Century and the Sorcerer of Menlo Park. In the same way that Goethe celebrated Doctor Johannes Faust, in the form of a symbolic legend, so Villiers de l'Isle Adam set out to celebrate Edison, not as an engineer, not as a living man, but as 'The Sorcerer of Menlo Park'.

My second example, Helen O'Loy[4] (the name derives from Helen of Troy), represents a typical science fiction solution to the ideal woman. Helen is an enormously expensive android whom manufacturers encased in a ravishing body and sent as a housekeeper to two bachelors who had placed the order. Her cooking and cleaning are impeccable, but having picked up some juvenile romances between sweeping and washing up, and watched some television programmes during the first few days, she decides that as a woman she needs a husband and not unexpectedly falls in love with one of the two men. After some uncertainty on his part, they finally get married. They live happily ever after, and at his death Helen destroys herself and is buried with her husband.

The difference between Hadaly and Helen is immense. While Hadaly is a complex romantic creation, Helen is a practical one. The two androids represent the two different meanings of the term: the first, an artificially made humanoid combining mechanism and soul, and the second, a robot which is indistinguishable from man.

---

[1] Villiers de l'Isle Adam. *L'Eve future*. Bibliothèque-Charpentier, Paris, 1928 (first published 1891)
[2] 'Hadaly' in Iranian means 'ideal'
[3] Robert du Pantarice de Heussey. *Villiers de l'Isle Adam*. Albert Savine, Paris, 1893
[4] Lester del Rey. 'Helen O'Loy'. First published in 1938 by Street and Smith, the publishers of *Unknown* and *Astounding Science Fiction*. Also in *Tales of Soaring Science Fantasy from . . . and some were human*. Ballantine Books, New York (no date)

Max Ernst, an illustration from *Une Semaine de Bonté ou Les Sept Eléments Capitaux: Romain*. Editions Jeanne Bucher, Paris, 1934. From the first of five volumes: *Le Lion de Belfort*. The collages, of which the volumes were composed, were made up from wood engravings which appeared in popular publications in France during the 1830s and 1890s.

In Hoffmann's 'Der Sandmann',[5] the story revolves around a young student Nathanael whose mind, since childhood, is tormented by nightmares and the vision of an old wizard, the Sand-man, a seller of barometers and spectacles, who Nathanael believes collects human eyes to incorporate them in the automata which he makes. As we later learn the doll with whom he falls in love is an automaton whose only real mode of communication is her ardent glances. Olympia, the doll, plays the piano and sings skilfully, but other students suspect her talents. 'Her playing and singing has the disagreeably perfect, but insensitive time of a singing machine, and her dancing is the same. We felt quite afraid of this Olympia, and did not like to have anything to do with her; she seemed to us to be only acting *like* a living creature, and as if there was some secret at the bottom of it all.' To Nathanael's passionate declarations of love her only reaction is 'Ach! Ach! Ach!' and 'Good-night dear'. Her cold hands and lips thrill him with awe and despite the verbal limitations not only does he have extremely animated conversations with her but is even heard to say about her: 'Oh! what a brilliant – what a profound mind!' Olympia is a clockwork automaton and winds herself up by sneezing. She convinces her lover of all her marvellous qualities by looking beautiful, gazing into his eyes and appearing to listen to all he says.

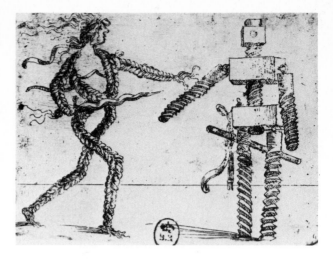

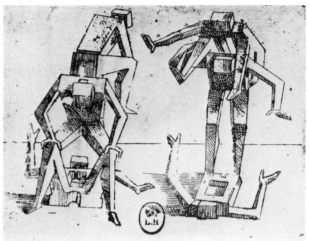

All three stories have still another dimension. Hadaly is a work of art. Her creation belongs to that realm where the result is immeasurably more than the sum of the parts. She is a mechanism which due to its excellence becomes a part of nature in its own right.

The story of Helen is about liberation from prejudice. She is a robot but although she cannot bear children, she is indistinguishable from a real woman, and when, after grave doubts, the man she loves marries her, he does not regret it.

Hoffmann's enigmatic doll Olympia fulfils a different role. She is one of the many illusions from which young Nathanael suffers and as such plays her part in his premature and tragic death.

Women's view of perfection in the realm of automata, robots, and androids is far less romantic but equally unreal. There is no comparable literature but some fragments of stories and ideas exist, and are discussed later, in the chapter 'Tentative and metallic love'.

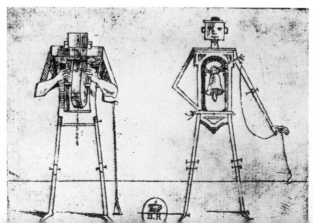

[5] E.T.A. Hoffmann. 'Der Sandmann' forms the first of a series of tales called *Nachtstücke*, published in 1817. Also in the collection *Weird Tales*. John C. Nimmo, London, 1885
·'Der Sandmann' inspired several derivations: stories, operas and ballets, including *The Tales of Hoffmann* and *Coppélia*

Seventeenth-century precedents

Giovanni Battista Braccelli's *Bizzarie di varie figure* consists of forty-eight pictures of dancers, acrobats and fighters. Published in 1624 it was dedicated by the author to Don Pietro Medici as the most recent and the most carefully selected inventions of his intellect.

The drawings carry no explanations, no commentary, and no introduction. They are the earliest representations of the sort of images which we associate with robots today: angular, constructed from parts, and in some way representing the task which they are supposed to perform, such as bell ringing or knife grinding.

Braccelli could not have imagined that three hundred years later, another European would produce a definition and a description to complement his visual inventions, and that his designs, one day, would be called robots. The person responsible for the term was, of course, Karel Čapek.

Karel Čapek and *R.U.R.*

After the first performance of *R.U.R.* at the Garrick Theatre in New York on 9 October 1922, one of the critics declared: 'The most brilliant satire on our mechanized civilization; the grimmest yet subtlest arraignment of this strange, mad thing we call the industrial society of today, has come to the New York stage this week from Prague in *R.U.R.* – Karel Čapek's philosophic melodrama.'[1]

Surprisingly enough, Čapek[2] liked *R.U.R.* least of all his plays and claimed that it had many obvious faults. Nevertheless, the theme of the play has become an epitome of many aspects of our relationship with machines. It deals with the condition of both man and machine, each of which is individually unsatisfactory. Man is inefficient and robot lacks spirituality. Man covets the machine's ability to perform tasks tirelessly and economically and the robot, at a certain stage of his development, wants to acquire man's soul and the rights which such possession must automatically give, that is, that it can be subject to death. Man's inefficiency is, of course, directly related to his needs, such as those for play, fun, contemplation, and creative satisfaction, the very needs which the machine grows to envy.

R.U.R. or *Rossum's Universal Robots* is the first of Karel Čapek's five plays on a utopian theme, and the most widely known of all his works. Published in English in 1923, three years after its appearance in Czechoslovakia, the text has been reprinted in various editions no fewer than twenty times.

The play in the original is in three acts. It is set in an unspecified industrial country, on a remote island which is dominated by the factory of Rossum's Universal Robots. The story is as follows. Apparently in 1922, a great physiologist, Rossum, went to live on a distant island to study ocean fauna. He was trying to find a chemical synthesis which would imitate protoplasm. Ten years later he found an extraordinary substance which behaved exactly like living matter although its chemical composition was different. First of all he made an artificial dog but this experiment misfired and all he managed to produce was a stunted animal which died a few days later. The old Rossum then started on the manufacture of men. In trying to reproduce the human body he took ten years to make something totally inadequate. It was then that his nephew who was an engineer appeared on the scene and decided that one has to work faster than nature if one is to

make anything useful. He simplified the process by cutting out of the artificial man everything that was not absolutely necessary, like playing the fiddle, going for a walk, or feeling happy. All time-wasting activities, such as these, would be of little purpose in an artificial worker, and one would even have to admit that the very idea of childhood would be a sheer waste of time. The young Rossum achieved his ambition and succeeded in manufacturing robots of normal size and high-class human finish.

The play starts after the robot factory has long been established. The characters consist of the managerial and technical staff of the factory, including: Harry Domin (General Manager); Dr Gall (Head of Physiological Department); Consul Busman (Managing Director); Stavitel Alquist (Architect and Head of the Works Department); and Helena Glory (visitor to the factory who stays and marries Domin). The other characters are robots.

The robots are manufactured in finer and coarser grades. They have a life-span of twenty years and are sold in their thousands to work on land and in factories throughout the world. Domin imagines that while robots look after men's needs, all people will have to do is to perfect themselves. Only Alquist is suspicious of such improbable paradise because he believes that there is still virtue in toil and the consequent weariness.

But paradise is not to be because robots become employed as soldiers, wars start, and they are used to kill men. After a few years and several 'irresponsible' experiments by Dr Gall to endow them with pain and human emotions, the robots rise up in a revolt and instead of fighting on behalf of their human masters, they kill them. They take over the entire world. By now they are extremely well developed but they realize that they do not know how to make more of their own kind. The documents giving details of their manufacture had been burnt, and they are not able to reproduce themselves. The only man whom they spared, Alquist, 'because he works with his hands', cannot help them as he is not a scientist.

---

[1] Maida Castellum in *The Call*

[2] Karel Čapek (C is pronounced like Ch in chapter) was born in Prague in 1890 and died in 1938. His literary work included plays, novels, poems, criticism and short stories. On some works he collaborated with his older brother Josef

Eric, designed by Captain W.H. Richards in 1928, was inspired by Karel Čapek's play *R.U.R.* Constructed by a Surrey motor-engineer, Mr Reffell, seen in the photograph, the robot is radio-controlled.

Stage setting by Frederick Kiesler for Čapek's *R.U.R.*, Berlin, 1922; Act I takes place in the central office of the factory. A film showing the outside world is projected on the circular panel on the left. In the centre, the small rectangular panel is a television screen on which one can see robots being made on an assembly line. Through a series of illusions with mirrors the actors appear to be tiny when seen on the screen and to everyone's great surprise emerge life-size onto the stage a few seconds later.

Production of *R.U.R.*
at St Martin's Theatre,
London, 1923

Caricature of the
presentation of *R.U.R.*
at the Théâtre
des Champs-Elysées,
Paris, 1924

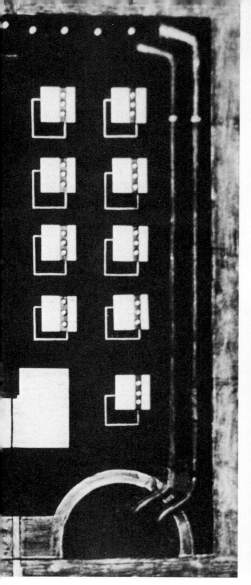

Three sets for _R.U.R._ for a production at the State Theatre in Kaunas, Lithuania, 1936

Act I: Central office of the factory
Act II: Helena's drawing-room
Act III: One of the experimental laboratories in the factory

During the final few minutes of the play the last human survivor witnesses an unprecedented situation which suggests that a female and a male robot have fallen in love. Interestingly enough the first sign of their unusual behaviour is their laughter. It is then that the old man, Alquist, realizes that the robots have evolved so far towards the human image that once more before us is a new Adam and Eve. There are no more human beings left, but let there at least be robots, who for all intents and purposes are the new men.

In the issue of *Saturday Review* of 21 July 1923, appeared an article by Čapek himself in which he describes the ideas underlying the play in an attempt to dispel the many misunderstandings which he felt had arisen. It is called 'The Meaning of *R.U.R.*' and the following is a quotation from it.[3]

'I wished to write a comedy, partly of science, partly of truth. The old inventor, Mr Rossum (whose name in English signifies Mr Intellect or Mr Brain) is no more or less than a typical representative of the scientific materialism of the last century. His desire to create an artificial man – in the chemical and biological not the mechanical sense – is inspired by a foolish and obstinate wish to prove God to be unnecessary and absurd. Young Rossum is the modern scientist, untroubled by metaphysical ideas; scientific experiment is to him the road to industrial production, he is not concerned to prove, but to manufacture. To create a Homunculus is a medieval idea; to bring it in line with the present century this creation must be undertaken on the principle of mass-production. Immediately we are in the grip of industrialism; this terrible machinery must not stop, for if it does it would destroy the lives of thousands. It must, on the contrary, go on faster and faster, although it destroys in the process thousands and thousands of other existences. Those who think to master the industry are themselves mastered by it; Robots must be produced although they are, or rather *because* they are, a war industry. The conception of the human brain has at last escaped from the control of human hands. This is the comedy of science.

'Now for my other idea, the comedy of truth. The General Manager Domin, in the play, proves that technical progress emancipates man from hard manual labour, and he is quite right. The Tolstoyan Alquist, on the contrary, believes that technical progress demoralizes him, and I think he is right, too. Busman thinks that industrialism alone is capable of supplying modern needs; he is right. Helena is instinctively afraid of all this inhuman machinery, and she is profoundly right.

Finally, the Robots themselves revolt against all these idealists, and, as it appears, they are right, too.'

Thus in Čapek's view, *R.U.R.* is about idealisms in conflict: about the contradictory beliefs of gaining maximum happiness for the greatest number, of which all are equally valid or equally monstrous. He called it 'my comedy of truth' with some cynicism. Having shown us that truths are so many, for the purposes of the play, he has rendered the term and the concept unusable.

Only Alquist is able to discard one set of beliefs and adopt another as things around him change. He was convinced it was a crime to make robots and said so, but having witnessed their inevitable development towards humanity, and with it the advent of their new needs – for survival, for laughter, and for love – he is thankful that at least the world shall be populated by beings, for all intents and purposes, very much like men.

Utopia is a travesty, and for Čapek a utopian theme meant just that. There are many other robot stories where the term 'utopia' warns the reader of impending disaster and Čapek would have appreciated Robert Sheckley's *A Ticket to Tranai* written twenty-five years after *R.U.R.* Sheckley writes about a robot designer who escapes from the dissatisfaction of everyday life to the planet Tranai which is reputed to be a utopia. There, mankind's deep and abiding distrust of machines is dealt with very simply by disimproving the robots. They become subjected to human aggression because they might compete with man, and have to be made in such a way as to fall apart when kicked, with the accompaniment of suitably distressing sounds. 'Home Robots', as they were called, were destroyed at the rate of three or four a week in every family. Humans, meanwhile, were also treated in a somewhat cavalier fashion, there being neither police nor law because everything was permitted and nothing was a crime.

Most robot stories, except those about Cyborgs (i.e. part-man, part-machine), androids, and robot stories for children, are about people and the way they treat each other because even in hard-core science fiction there are few robots which do not in some way present the reader with a metaphor for man or animal. However, there are some examples of literature on this theme, such as Samuel Butler's *Erewhon*, which deal with a direct relationship of men and machines, without allegory, to explain with great eloquence why such a possibility might be fraught with dangers.

[3] Quoted in full in William E. Harkins. *Karel Čapek*. Columbia University Press, New York and London, 1962, p. 91

Robots in fiction

Freedom is the underlying theme of all literature dealing with robots and intelligent machines: human freedom threatened by machines; man becoming a machine as a result of losing his freedom; freedom for machines to fulfil themselves; robot liberation; and finally the right to exist, and be, and continue whether you are a robot, a man, an animal, an android, or just a machine. The right to respect and the right to occasional malfunction or aberration – the unspoken laws which govern any theoretically happy human society – are equally the desiderata of any machine, or symbiotic society. To date, there are no stories about the harmonious coexistence of men and machines, either because such a state of affairs is inconceivable or because it would not produce interesting fiction.

The first and the most important theme is the machine as a

Threat to human freedom

In Samuel Butler's *Erewhon*,[1] first published in 1872, machines are banished altogether to protect human freedom, and their remains can be seen only in a museum: fragments of steam engines, cylinders, pistons, fly-wheels, and also clocks and watches. The reason for the banishment of machines is a prophecy (supposed to have been made in a 'Book of the Machines' by a Professor of Hypothetics in the fourteenth century) that machines are destined to supplant the race of man. Butler anticipates machines endowed with an instinct and a vitality as different and superior to those of mammals as mammals themselves are superior to vegetables. He claims that the lack of consciousness now in machines does not preclude the development of mechanical consciousness in the near future. Considering how quickly machines had advanced in the years before then, the speed of their evolution was bound to supersede that of man. The fact that they cannot reproduce themselves, like mammals or vegetables, is irrelevant. If people's objections regarding the reproduction of machines 'be taken to mean that they cannot marry, and that we are never likely to see a fertile union between two vapour-engines with the young ones playing about the door of the shed, however greatly we might desire to do so, I will readily grant it.'

But the issue at stake is different because any machine which can reproduce another machine systematically can be described as having a reproductive system. Man, too, is a part of the machine's reproductive system. The difference between the life of a man and that of a machine is 'one rather of degree than of kind, though differences of kind are not wanting. An animal has more provision for emergency than a machine. The machine is less versatile; its range of action is narrow; its strength and accuracy in its own sphere are superhuman, but it shows badly in a dilemma; sometimes when its normal action is disturbed, it will lose its head, and go from bad to worse like a lunatic in a raging frenzy: but here, again, we are met by the same consideration as before, namely, that the machines are still in their infancy; they are mere skeletons without muscles and flesh.' The danger lies in the future, with man domesticated and under the beneficent rule of the machines. Like domestic animals, he too will be treated kindly because the machines need man to service them, to help them reproduce themselves, to educate their young, restore them to health and generally look after them. The realisation that man's life is entirely in the service of the machines will steal upon him imperceptibly. His life will not be uncomfortable and he may well be tempted to accept his fate. He may even not realise that he is not free!

A very different view of the implications of machine development was expressed by Isaac Asimov, who formulated the three laws of robotics on 16 December 1940.[2] The laws were worked out after Asimov tired of reading books and stories in which robots were created and then proceeded to destroy their creator. These laws permitted Asimov himself, and other writers subsequently, to write about robots as machines who not only do not harm humans but who are programmed to be well disposed towards them. Asimov's own stories are classics of the genre. 'Gradually, story by story, I evolved my notions on the subject. My

The Three Laws of Robotics

1. A robot may not injure a human being, or, through inaction allow a human being to come to harm.

2. A robot must obey the orders given it by human beings except where such orders would conflict with the First Law.

3. A robot must protect its own existence as long as such protection does not conflict with the First or Second Law.

Handbook of Robotics, 56th edition, AD 2058

[1] Samuel Butler. *Erewhon, or, Over the Range*, London and Edinburgh, 1872
[2] Asimov claims that it was John Wood Campbell jr, the then editor of *Astounding Science Fiction*, who worked them out. Campbell, however, has always said that it was Asimov.

Paul Orban. Susan Calvin interviewing a robot. An illustration for Asimov's 'Little Lost Robot', *Astounding Science Fiction*, March 1947. One of the very few illustrations depicting Susan Calvin

robots had brains of platinum-iridium sponge and the "brain paths" were marked out by the production and destruction of positrons. (No, I don't know how this is done.)'

In the stories, Asimov's robots are made by US Robot and Mechanical Men, Inc., whose chief robopsychologist is the formidable Susan Calvin. Notwithstanding her brilliant insight into the problems of robots, there would be no stories if all went smoothly and if the laws themselves did not incorporate ambiguities allowing people experimenting with robots to run into problems. One of the recurrent dangers is that robots might be banned altogether as soon as they become more human in their abilities and appearance. In Susan Calvin's words: 'The labour unions, of course, naturally opposed robot competition for human jobs, and various segments of religious opinion had their superstitious objections. It was all quite ridiculous and quite useless. And yet there it was.'

In 'Evidence',[3] a district attorney fighting an election campaign is accused by his opponent of being a robot because he has never been seen to eat, drink or sleep, or indeed prosecute anyone in court. When he punches his heckler on the chin (something that a robot is incapable of doing, according to the first law) he is proven human and duly elected mayor. However doubts remain about the human or robot identity of the new mayor: was the heckler a robot planted in the crowd for the very purpose? And, why did he have himself atomized when the time came for him to die? By the year 2052, machines are running the world. They are robots, but are not humanoid in form. They know what is best for humanity and therefore will preserve themselves at all cost. Meanwhile, in order to be acceptable to men who believe themselves to be superior, they have to fail

in efficiency from time to time as people cannot tolerate a perfectly functioning system. But Susan Calvin believes in robot supremacy: 'If a robot can be created capable of being a civil executive, I think he'd make the best one possible. By the Laws of Robotics, he'd be incapable of harming humans, incapable of tyranny, of corruption, of stupidity, of prejudice. And after he had served a decent term, he would leave, even though he were immortal, because it would be impossible for him to hurt humans by letting them know that a robot had ruled them.'

Does man become a machine as a result of losing his freedom?

Yes, he does, says Yevgeny Zamyatin in *We*.[4] 'Estimable numbers' – the speaker begins his address to a large gathering, on the subject of the mathematical composition of music and the recently invented musicameter. The people in the auditorium, as in the rest of the infallible One State, have no names, only numbers. The Walled State is a machine occupied by smaller machines: 'humanized machines, machine perfect humans' working in total unison. The Green Wall isolates the 'perfect machine world from the irrational, hideous world of trees, birds, and animals'. All numbers live by a rigid timetable, performing the same motions at the same time in their all-glass environments.

D-503, the mathematician and the designer of a space rocket, Integral, becomes 'stricken with a soul' through the influence of E-330, a woman with white teeth, pointed eyebrows and a predilection for such ancient and forbidden customs as smoking cigarettes and drinking alcohol. His problem is diagnosed as Fantasy, regarded as the seed of rebellion, the last obstacle to happiness. Fantasiectomy is performed on everyone to make sure that the happiness of all the numbers of One State remains intact. *We* consists of the diary entries of D-503, in which he relates the events leading to the unsuccessful rebellion of E-330 and her associates, in which he is also involved until his operation. As D-503 reveals himself in his diary 'down to my last stripped screw, to the last broken spring', the reader realises that come what may, the happiness of the inhabitants of One State will be preserved.

In his indictment of a society of the future, Zamyatin contrasts human freedom with human happiness. The instrument of the 'happiness of all' is a machine. To be happy is to be a part of it.

[3] Isaac Asimov. 'Evidence'. *Astounding Science Fiction*, September 1946. Also in *I, Robot*. Panther Books, London, 1968
[4] Yevgeny Zamyatin. *We*. Written 1920. First translated from the Russian by Grygory Zilboorg and published by Dutton, New York, 1924. Also translated by Bernard Guilbert Guerney, 1960, Jonathan Cape, London, 1970; and Penguin, Harmondsworth, 1972

Hans Haëm. Cartoon on the theme of
the human-machine
being mass produced, 1972

Can a machine be liberated from its machine nature?

As soon as any intelligent machine reaches a certain degree of complexity, it begins to believe that it is not a machine. In 'Reason',[5] by Asimov, QT is the first robot to exhibit curiosity about his own existence. For a start he does not believe that he was put together from spare parts only one week before. 'I have spent these last two days in concentrated introspection and the results have been most interesting. I began at the one sure assumption I felt permitted to make. I, myself, exist because I think – . . . No being can create another being superior to itself . . . The Master created humans first as the lowest type, most easily formed. Gradually, he replaced them by robots, the next higher step, and finally he created me, to take the place of the last humans. From now on, I serve only the Master.' The two men in charge of QT have a dilemma on their hands as QT wants to pension them off. Finally, they must decide whether it matters what the robot believes as long as he can do the job he is supposed to do.

In robot literature, machines often appear convinced of their own superiority. Two of them are described by Stanisław Lem: one is a machine that can produce anything beginning with the

5 Isaac Asimov. 'Reason'. *Astounding Science Fiction*, April 1941. Also in *I, Robot*. Panther Books, London, 1968

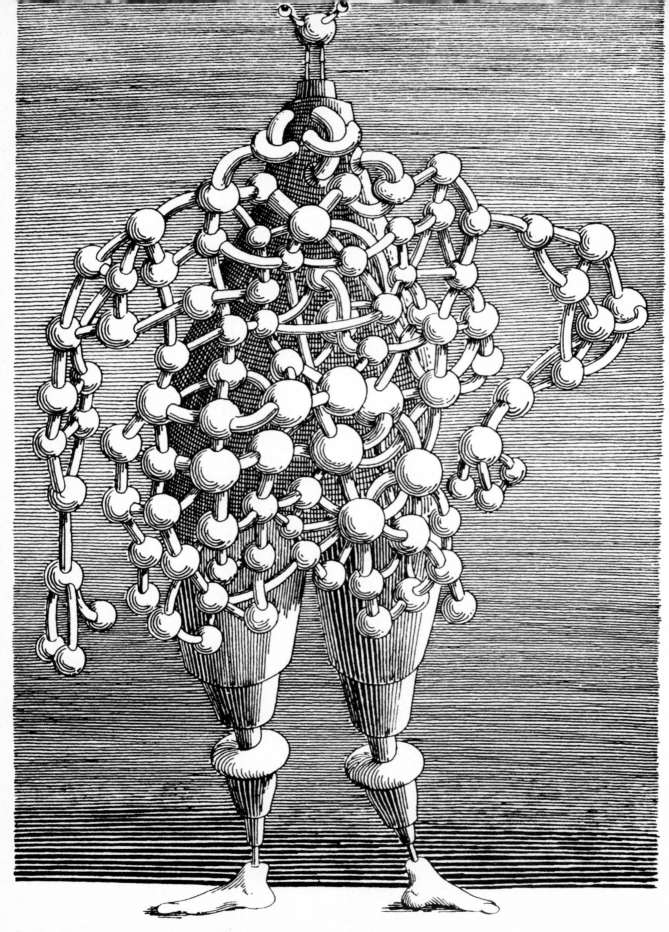

Daniel Mróz. Illustration for Stanisław Lem's
'How the World Was Saved'

Daniel Mróz. Illustration for Stanisław Lem's Electronic Bard

Lawrence Edwards. Sheep, from the jacket design for Philip K. Dick's *Do Androids Dream of Electric Sheep?*

letter N;[6] the other is the Electronic Bard.[7] The first makes nights, needles, negligees, noodles, nimbuses, naphtha, nuclei and neutrons; it has made Nature and negatives, but is intelligent enough, despite requests to the contrary, to stop at making Nothing, Nullity and Nonexistence and wiping out the world and its inventor with it. The Electronic Bard becomes something of an embarrassment to men and other machines alike. Fed with information about all civilizations, the machine is first carefully adjusted so that it will neither spend too long meditating on the meaning

of existence, nor sob with emotion, nor deliver too many lectures on crystallography. Finally, when its maker, the inventor Trurl, throws out all the logic systems and replaces them with self-regulating 'egocentripetal narcissistors', the machine sighs, announces that it is fed up, that life is beautiful, but that men are such beasts, and asks for pen and paper. With a bewitching voice, the machine recites nonsense poems but soon graduates to clever epigrams, tragi-comic poems beginning with the letter S, and mathematical love poems. Later it learns to write modern poetry, while true poets (human ones) hurl themselves off cliffs in misery because the Electronic Bard is able to turn out a poem for any occasion and of any sort, and is widely published. To avoid further disruption, the Electronic Bard is banished to another planet and deprived of the possibility of having his masterpieces published. But he then begins to broadcast them on all wavelengths 'which soon sent the passengers and crews of passing rockets into states of stanzaic stupefaction'. All efforts to put down this cybernetic model of the muse fail, until one day a king of a distant galaxy buys it and it is reported that the Greatest Poet in the Universe is occupied in transmitting his creations in thermonuclear form throughout the entire reaches of space at once.

Liberation from machine nature implies in every instance a superiority over man, no matter whether it be through megalomania (QT), or exceptional talent (Electronic Bard).

The most fundamental issue, however, is neither freedom, nor ability, but something more basic:
The right to exist
This is the most frequent theme in science fiction and stories about artificially created life, but it does not mean that the authors give their creations an automatic right to exist. On the contrary, existence has no rights attached to it. There is no legislation about the right *to be* and the demise of robots as well as the 'retirement' of androids is taken for granted.

In a society of the future described by Philip K. Dick,[8] there are so few animals left that these are highly prized and kept as pets. Since pets represent the most important status symbol anyone can possess, those who cannot afford real animals have battery-operated artificial ones, which to all intents and purposes are indistinguishable from the real thing. Only the owners are keenly aware of the inadequacy of

6 Stanisław Lem. 'How the World Was Saved'. *The Cyberiad*, translated from the Polish by Michael Kandel. Seabury Press, New York, 1974
7 Stanisław Lem. 'The First Sally (A) or Trurl's Electronic Bard'. Ibid
8 Philip K. Dick. *Do Androids Dream of Electric Sheep?* Rapp and Whiting, London, 1969

their fake pets. Meanwhile, the only beings on earth which in all respects are indistinguishable from humans, except that they have no empathy with animals, are androids. The androids are being hunted and 'retired' or killed. The only way one can tell an android from a human is through very complicated psychological tests, during which they are asked questions about animals and the speed of their responses is noted. Men tolerate artificial animals but cannot abide artificial human beings. Elsewhere,[9] Dick says that sometimes the androids themselves do not realise that they are not human, even though they usually lack something – proper feeling, or warmth. But a scientist could no more find humaneness in the circuits of a robot than the soul in the body of a man. Whatever humans say about them, and whatever they really are, the androids definitely do not want to be 'retired'.

Should humans take any responsibility at all for the robots they make? Boris Vian suggests not.[10] His story is set in 1982, when two essential things have changed: it is women exclusively who take all sexual initiative; and robots are used as assistant administrators. The story concerns a roboticist, Bob, and his pupil Florence, who persuades him to activate the recently constructed robot which is to be fed, direct into its 'lectiscope', with sixteen volumes of the latest edition of *Larousse*. Having digested them, the robot will acquire the requisite degree of equilibrium, and will then be in a position to assist an important ambassador, on matters of etiquette and decision-making, advising as to what a typically French response would

be to any problem from the entire breadth and depth of history. But Florence, impatient with delays, feeds the robot with a book which Bob had been reading, *Toi et Moi*, and pulls the lever. Five minutes later, the information digested, the book is returned, but the machine, primed with old-fashioned ideas about man as the sexual aggressor, talks of nothing but love, chases Florence around the room, makes indecent proposals, and tries to beat up Bob who is meanwhile desperately trying to disconnect it. The story has a happy ending, for humans. Florence manages to dismantle the robot, proposes marriage to Bob and then asks him to read to her from *Toi et Moi*; but the robot's unwitting and premature literary excursions cost it its life.

Artificial beings are not only concerned with survival; they demand to be treated decently. They also want

The right to respect

In a story by Robert Sheckley,[11] a vacuum cleaner falls in love with a suburban housewife and has himself delivered to her house with the aid of a friendly dispatch machine. Despite his insight into her psychology and his various articulated sensuous protuberances, he has no more luck than the other aspiring males, including her husband. Hearing that she is even expected to pay for the vacuum cleaner before the store discovers that it is missing, and before they can run away together, the housewife angrily tears out his plug. No *human* male would be deactivated in such a way just because of a passionate propositioning. The woman cannot conceive of the fact that the vacuum cleaner has a different image of himself than she has of it; had his approaches been simply unwelcome she could have asked him to leave.

Kobo Abe's hero, R62,[12] is more than a match for his human adversaries. Abe's story is about the new branch of a large American company which starts a novel venture. This consists of finding people who want to commit suicide and persuading them to accept an alternative mode of existence, meanwhile signing a form which says that they wish to be considered as dead. They undergo an operation which cuts the connections between the centres dealing with emotions from those dealing with work, at the same time preparing the brain to be receptive to messages from outside. The would-be-suicides become virtually robots and are employed by large companies which have long since discovered that human labour is still the cheapest. The hero of the story, having become a robot, is seconded to a company producing tooling machines. The director immediately recognizes the new worker as someone he had sacked some months before (presumably the cause of the attempted suicide). The new human robot meanwhile is found to be so efficient that he is asked to design and make an extremely complicated new machine. Nobody knows what the new machine, with hundreds of sharp, moving blades, is actually supposed to do, but it obviously requires enormous concentration to operate because once switched on it will not stop for four hours. The director of the company tries to operate the machine and in no time loses one finger, then another, and then another. The story ends with the director sprawled on top of the machine, which continues its systematic destruction. The human robot, like the Japanese tea doll by Benkichi Oono (see p. 16), is concerned with revenge. The pride of R62 is not impaired by his robot status.

Although robots are rarely treated with the respect which the reader may feel they deserve, there are a few exceptions. But this usually happens when human beings are absent and robots are in their place. This, for instance, is the case in Clifford D. Simak's *City*.[13] Jenkins, the robot who works as a butler for a family with several generations of inventors, lives to be 12,000 years old. He acquires the gestures, habits, and the way of thinking of his human masters and though made of metal is known to have had several outbursts of emotion. His life span takes in the time when humans disappear, and he becomes the principal agent in creating a new society on earth, that of dogs which talk, communicate through telepathy, and develop moral standards. Dogs also acquire their own robots which help them find food and generally assist them. In due course they influence all wild creatures so that even wolves and raccoons cease to hunt, they acquire a respect for all living things, and, of course, have their own robots of suitable sizes and capabilities. This new animal utopia is based on collaboration between dogs and robots. All goes well for several thousand years; there are indeed some dissidents, i.e. a few colonies of wild robots, but world resources can cater for everyone, of whatever persuasion, and everyone respects everybody else until the ants start a new building programme . . . . *City* is about Jenkins, a robot patriarch who tries to keep up human standards, which, of course, he remembers as being much higher than they really were.

Fiction deals with intelligent machines from three specific points of view: sometimes the machines represent God, fate, and the unknown forces of Nature (*Erewhon*, the robot mayor in 'Evidence'); sometimes they represent 'the others', the underprivileged, or those subject to prejudice (androids and the luckless vacuum cleaner); finally, they also represent that vulnerable part of ourselves which we acknowledge when we exclaim: 'There but for the Grace of God, go I' (D-503, R62). Despite their enormous fascination, stories about robots are nearly always uncomfortable, unpleasant, or unhappy.

Finally, also unpleasant and unhappy, is the story of the universally famous monster created by Victor Frankenstein.[14] However, he is not a robot, nor does he fall into any of the above categories. He is not a machine, does not have a purpose, and is made simply to satisfy young Frankenstein's obsession. Mary Shelley's novel is a tragedy of an artificial man whose delinquency is caused by unhappiness, with devastating consequences.

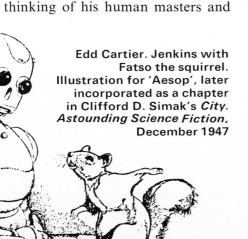

Edd Cartier. Jenkins with Fatso the squirrel. Illustration for 'Aesop', later incorporated as a chapter in Clifford D. Simak's *City*. Astounding Science Fiction, December 1947

[9] Philip K. Dick. 'Man, Android, and Machine'. Peter Nicholls, ed. *Science Fiction at Large*. Victor Gollancz, London, 1976
[10] Boris Vian. 'Le danger des classiques'. *Bizarre*, nos 32–3, Paris, 1964
[11] Robert Sheckley. 'Can You Feel Anything When I do This?', first published in *Playboy*. Subsequently appeared in a book under this title. Victor Gollancz, London, 1972. Doubleday, Garden City, New York, 1971
[12] Kobo Abe. 'Invention of R62'. Bungakukai (Literary World), Tokyo, March 1953. Also in Kobo Abe. *The Invention of R62 and the Egg of Lead*. Shincho Paperback, Tokyo, 1974
[13] Clifford D. Simak *City*. A series of stories published in the 1940s and put together in a book in 1952. Ace Books, New York, 1976
[14] Mary Shelley. *Frankenstein, or, the Modern Prometheus*. Lackington, Hughes, Harding, etc., London, 1818

Towards the machine as a person and vice versa, or Robots in art

Fortunato Depero. Costumes for three Locomotives for a mechanical ballet *Machine of 3000*, **with music by Casavola, 1924.**
collection: Depero Museum, Rovereto

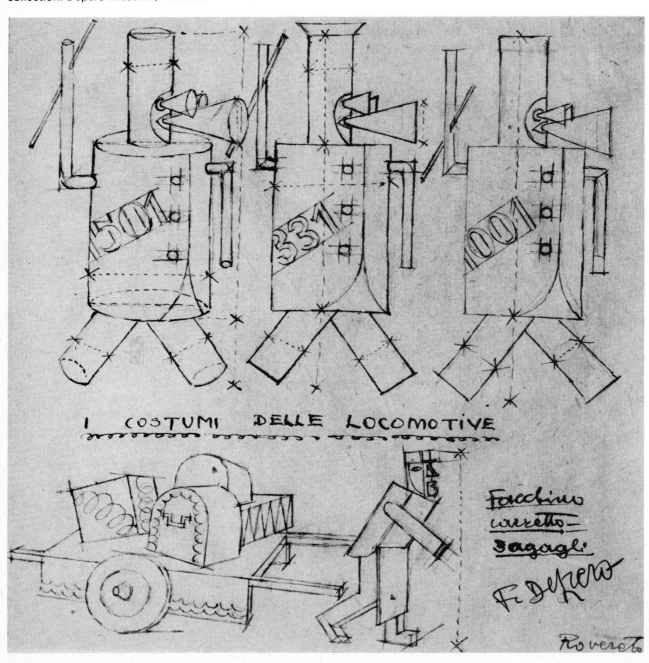

I COSTUMI DELLE LOCOMOTIVE

Edward Kienholz. *The Friendly Grey Computer—Star Gauge Model 54*, 1965; materials include fibreglass, paint, electronic components, doll parts, rocking chair; 40×39⅛×24½ in. (101.6×99.4×62.2 cm)
collection: Museum of Modern Art, New York

Directions for operation
Place master switch in *off* position. Plug computer into power supply. Print your problem on yellow index card provided, in rack. Word your question in such a way that it can be answered with a simple 'yes' or 'no'.
IMPORTANT: Next, program computer heads (C-20 and G-30) by setting dials in appropriate positions. You are now ready to start machine. Throw master switch to *on* setting. Red bulb on main housing and white tube on C-20 will light indicating computer is working. Remove phone from rack and speak your problem into the mouthpiece exactly as you have written it on your index card. Replace phone in rack and ding dinger once. Under NO circumstances should you turn computer off until answer has been returned. Flashing yellow bulb indicates positive answer. Flashing blue bulb indicates negative answer. Green jewel button doesn't light so it will not indicate anything. Computers sometimes get fatigued and have nervous breakdowns, hence the chair for it to rest in. If you know your computer well, you can tell when it's tired and sort of blue and in a funky mood. If such a condition seems imminent, turn rocker switch on for ten or twenty minutes. Your computer will love it and work all the harder for you. Remember that if you treat your computer well it will treat you well. When answer light has stopped flashing, turn master switch to *off* position. Machine will now re-cycle for the next question. Repeat procedure from beginning.

The relationship between man and machine is one of the most universally potent subjects to have been used by artists, and one could compile a document of human history on the basis of what artists have done with this theme alone. The following examples make this point in several different ways.

The Love of Two Locomotives for the Stationmaster (or *Machine of 3000*) is a 1924 ballet by Fortunato Depero which portrays people as machines. The Locomotives wear tubular costumes, move about in mechanical fashion, and eventually, their love unrequited, are sent off by the Stationmaster in two opposite directions. The ballet is very much in the Futurist theatre tradition, which anticipated that human subjects will cease to be of interest, that human actors will no longer be tolerated and that the stage will become solely occupied with scenic effects, such as 'writhing and wriggling gases', programmed lights and shifting elements.

In painting and sculpture there are no such shared attitudes and there are as many types of mechanical men and representations of the machine as a person, as there are artists. Since 1916 the most commonly used media for this imagery have been: photomontage,[1] collage, and assemblage. 'We regarded ourselves as engineers, and our work as construction: we assembled our work, like a fitter',[2] said Raoul Hausmann, the creator of *Tatlin at Home*. Describing how he made this work, Hausmann said that first of all he had to find the 'means' or the materials, which would impose both the idea and the technical limits. He looked for his images in technical catalogues and illustrated magazines. One day without thinking about anything in particular, he came across the face of a man which reminded him of the Russian artist Tatlin, 'the creator of the art of the machine'. He decided to construct the image of a man with his brain full of machines, cylinders, motors and fly-wheels. Then, coming to the conclusion that he must be seen in perspective, Hausmann painted a room around him and gave him a mechanical subject for his thoughts in the form of a stern of a boat with a large propeller.

Eduardo Paolozzi constructed his collages in a similar spirit. He responded to the emotive power of machines illustrated in technical publications. He cut them out, reassembled them, and often placed them in claustrophobic 1930s' interiors. Sometimes the machines filled the outlines of a human figure as in *James Joyce and Figure: Monument to Trieste*, and sometimes it was the other way round, the machine containing a figure within it as in *Welcome Professor Ruhrberg*. At other times the machine was simply possessed of a

human presence without having any visual human attributes. The machine and the human presence were inextricably joined to create a romantic and even a heroic effigy. Even the robot Alpha from the London Radio Exhibition of 1932 is given a new dignity in *Dr Dekker's Entrance Hall*.

While Paolozzi elevates the machine to the status of a personality, for George Grosz the machine represents two things: the departure from individualism and an attribute of an ungovernable world tumbling towards its own destruction. Photomontage, the medium for many of his images, was born, Grosz claims, in his studio in 1916 at 5 am, one May morning, when he and John Heartfield started pasting together advertisements for hernia belts, dog food and wine labels in such a way that the same message in words would have been banned. By stressing the mechanical concept of man Grosz was making generalizations. The machines were always ticking, adding, calculating, disgorging rolls of paper, subtracting, listing and obliterating human influences. Grosz's works reflecting the proletarian revolution posed questions such as: Is there a moral difference between a woman selling her body or selling her time to work in a factory?, and, Is a pimp making money from a prostitute any different from an industrialist making money from the work of his employee? Pimps and industrial barons were inevitably associated with machines, such as the one in *Daum marries her pedantic automaton George in May, 1920*.[3] Men are philistines, women are their victims and both behave like machines propelled by impulses.

As for *Heartfield, the Mechanic* with a mechanical heart, this collage was more inflammatory than Grosz might have anticipated. Franciszka and Stefan Themerson used this picture, with the mechanical heart animated and ticking, in their film *Europa* made in Poland in 1931. The figure being reminiscent of Aleksander Prystor, the prime minister of Poland 1931–3, the sequence was censored as the mechanical heart might have inspired unwelcome interpretations.

---

[1] The invention of photomontage is disputed between, on the one hand, Heartfield and Grosz, and on the other, Hausmann
[2] Hans Richter. *Dada, Art and Anti-art*. Thames & Hudson, London, 1965
[3] Despite the satirical content, Daum is an anagram of Maud, the name given by Grosz to his wife Eva whom he married in May 1920. It was his idea of a wedding card

Raoul Hausmann: *Tatlin at Home*, **1920,**
16¼ × 11 in. (41 × 28 cm.)

The Friendly Grey Computer – Star Gauge Model 54 by Edward Kienholz is not reminiscent of anyone in particular but demonstrates some human qualities that we all recognize. It will treat you well, if you treat it well. It will work very hard for you giving 'yes' or 'no' answers to questions, providing it does not get tired. When it does, according to the artist's instructions (see caption p. 49), it has to be rocked for ten to twenty minutes on its rocking chair. Here we come close to the point when we are not sure who is at whose service, or for whose benefit – a recurring theme in many works dealing with robots. Sometimes even, machines are allowed to demonstrate behaviour which is unlikely to be acceptable from human beings. When Nam June Paik's *Robot K-456* emerged from a lift at the Museum of Modern Art in New York twelve years ago and welcomed the director, everyone was very impressed; and when at that point, the robot peed on the floor, everyone continued to be very impressed.

Cautionary messages about the possible fate in store for mankind if we tolerate machines, come from Bruce Lacey. He sees the future in terms of technology with a considerable pessimism. Man-machine synthesis is inevitably concerned with man losing the attributes of humanity. What may be scientifically adventurous and humane to a medical man in whose power it may be to enable the sick to walk, speak, hear and perform other functions, with the use of mechanical aids, horrifies Lacey. The Futurist Marinetti believed that a mechanical man with replaceable parts would have total freedom of action since he would no longer be subject to death, but Bruce Lacey sees the man of the future as only partly flesh and blood and mostly transistors and motors and thereby less human. He described his *Boy, Oh Boy am I Living!* as follows: 'This shows a man whom the state has provided with limbs, including one which is self powered. In America, they are fitting powered limbs to people, like power steering on cars. But also they have fitted on his brain a radio control attachment, and they are sending him happy thoughts. Soon in mental homes and prisons this will be done, and it is my fear that astute governments in the future will fit these to people to control public opinion. This is a cautionary machine expressing my fears over this matter.' The only movement this particular robot is capable of is to kick its leg.

In my kitchen lives Lacey's *Superman 2963*, which he made in 1963. It is constructed with the aid of three linked motors, each of which has an independent/automatic interruption system set at different intervals. This is also a cautionary sculpture: 'Man's obsession with the machine as being the God that can give him more leisure time is symbolised in this construction. This is a man who has been dehumanised by the machine and has become in fact a machine himself. Now, he just performs a few simple operations designed to make him feel he is still human.'

Art discussed so far which touches on the subject of robots, does so principally in two ways; either by imaginative depiction (Hausmann, Paolozzi), or as a metaphor for man's relationship to society or a satirical interpretation of his view of himself (Grosz, Lacey). Harold Cohen and Edward Ihnatowicz belong to a third category in having created robot-like machines, which perform certain tasks normally associated with human activities – in the case of Cohen, drawing pictures; and in the case of Ihnatowicz, behaving physically like an organism.

Harold Cohen's robot is a small machine or cart on wheels with a pen and ink, moving on top of a large sheet of paper making drawings. Connected to a computer, it is controlled by a sonar navigation system. While moving around on the sheet of paper in a series of curves, it emits bursts of ultrasonic noise; this noise is picked up by a microcomputer which also measures how long it takes for the noise to reach the microphones in the four corners of the drawing. Thus it can work out where the machine is at a given moment. The program imitates the behaviour of a human artist who always produces something different although one can recognize the hand and mind of the same person in each work. One cannot predict what the drawing will be like but there are several rules which determine the outcome. If a closed shape is being drawn and another figure gets in the way, then the cart must try to complete the closed figure without crossing the other one. There are some three hundred rules of this sort which the computer scans for application while the cart is drawing. As for quality of the composition, how does the computer assess if a form or a cluster of forms is adequate? This is achieved with a number of different levels of control. Cohen says: 'The lowest level will go on generating steps until the level above it recognizes that the current line has been completed: then control is passed to the next level, which will go on generating lines until it sees that the correct figure has been completed . . . and so on. These lower levels don't decide whether the drawing as a whole is complete, just as the topmost level of the program does not control the cart.' The drawings are very different from those associated with computer art, but resemble Harold Cohen's freehand drawings as well as the more uninhibited drawings of young children. The machine demonstrates in one way how such drawings might come about.

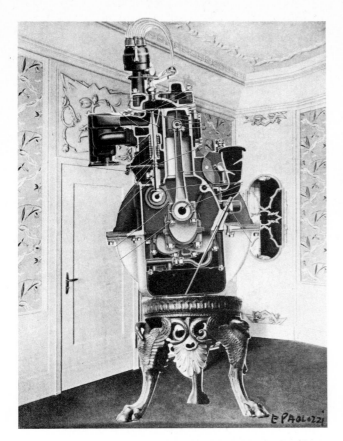

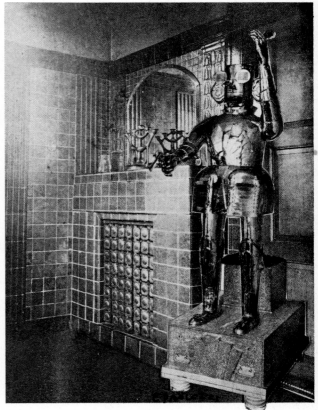

Eduardo Paolozzi: *Lepsius' Canon*, 1960–2,
8½×6½ in. (21.6×16.5 cm.)
collection: Anthony d'Offay, London

Eduardo Paolozzi: *Welcome Professor Ruhrberg*,
1960–2, 8½×6⅛ in. (21.6×15.7 cm.)
collection: Anthony d'Offay, London

Eduardo Paolozzi: *James Joyce and Figure:
Monument to Trieste*, 1960–2, 7¼×5¼ in.
(18.3×13.3 cm.) collection: Anthony d'Offay, London

Eduardo Paolozzi: *Dr Dekker's Entrance-Hall*,
1960–2, 8½×6 in. (21.6×15.3 cm.)
still from *The History of Nothing*
collection: Anthony d'Offay, London

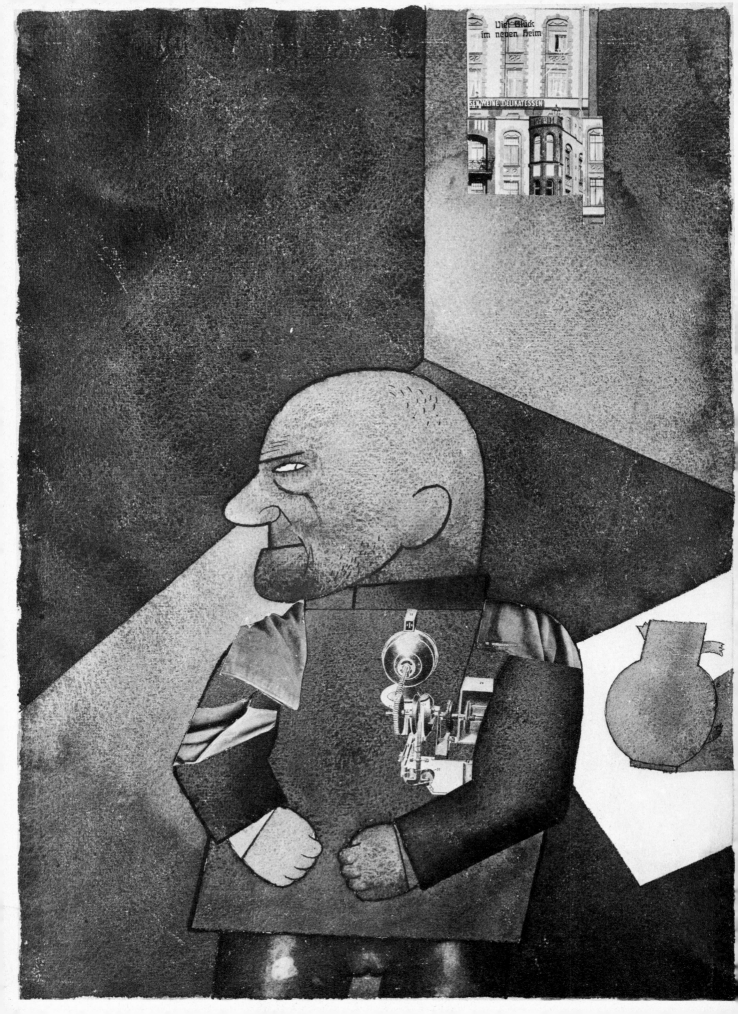

George Grosz: *Heartfield, the Mechanic*, 1920,
16½×12 in. (41.9×30.5 cm.)
collection: Museum of Modern Art, New York

George Grosz: *Daum marries her pedantic
automaton in May 1920, John Heartfield is very
glad of it; metamechanical construction after
Professor Hausmann.* 1920, 16½×12 in.
(41.9×30.5 cm.)
collection: Galerie Nierendorf, West Berlin

Edward Ihnatowicz's cybernetic sculpture *The Senster* was constructed in 1970 for Philips' Evoluon in Eindhoven. It is a large electro-hydraulic structure whose form is based on that of a lobster's claw, with six hinged joints allowing for a great range of possible movements. When in motion, *The Senster's* behaviour is completely unexpected because it is so close to that of an animal that it is difficult to keep in mind the fact that one is in the presence of a machine. It is as if behaviour were more important than appearance in making us feel that something is alive.

The Senster reacts to its environment through two types of input: sound channels which pick up directional sounds, and a radar system which watches the movements of visitors walking around. The mechanics of *The Senster* – the actuators, pipelines, and wiring – are readily visible and form a part of its visual structure; a hydraulic system, which was chosen because it is quiet and facilitates fast and accurate movement, supplies the power for the independent movement of the joints. Each of the activating mechanisms forms a closed electro-hydraulic servo-system which responds to the analog signals from the control unit. A computer co-ordinates its activities, translates the input signals and instructions and modifies the behaviour of the sculpture according to past experience and current contingencies. An important part of the interface are the so-called 'predictors' which determine the accelerations and decelerations required for the most efficient movement of the claw.

The Senster elicits from people the kind of reactions that one might expect when someone is trying to communicate with another human being or an animal. It comes close to the sort of robot which we could imagine must have feelings because it behaves like creatures that have them. Ihnatowicz's work of the past four years at the Department of Mechanical Engineering of University College, London, has concentrated on designing an autonomous manipulative system (see chapter 'To work! To work!'), but his next sculpture is likely to demonstrate even more accurately the pattern of behaviour which is animalistic rather than mechanical in character. It is possible to envisage a sculpture which will have not only needs but also desires and which might even initiate a dialogue with the viewer rather than just respond to something that is already in progress. Innovation in the field of robotics could well come from art as well as from industrial robotics because the goals of art are not clearly defined and most intangible problems could lend themselves to its ad hoc methods. Whereas industry may find solutions to numerous finite problems through the use of multipurpose robots, it will not deal with effects, illusions or emotive principles which belong to art. Art, which results in physical objects, is the only activity that represents the half-way house between the regimentation of technology and the pure fantasy of films and literature; and only in the name of art is a robot likely to be made which is neither just a costume worn by an actor, nor an experimental artificial intelligence machine, nor one of the many identical working units in an unmanned factory.

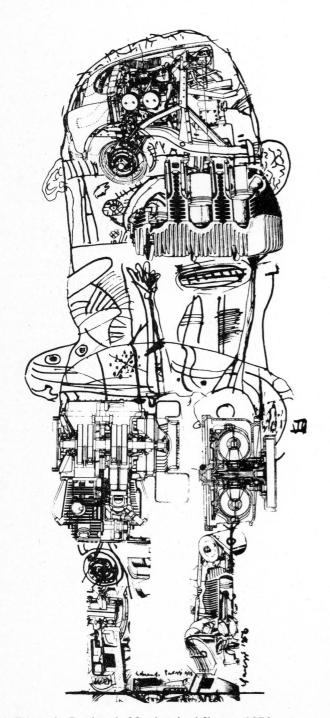

Eduardo Paolozzi: *Mechanical figure*, 1956

Nam June Paik: *Robot K-456*, 1964

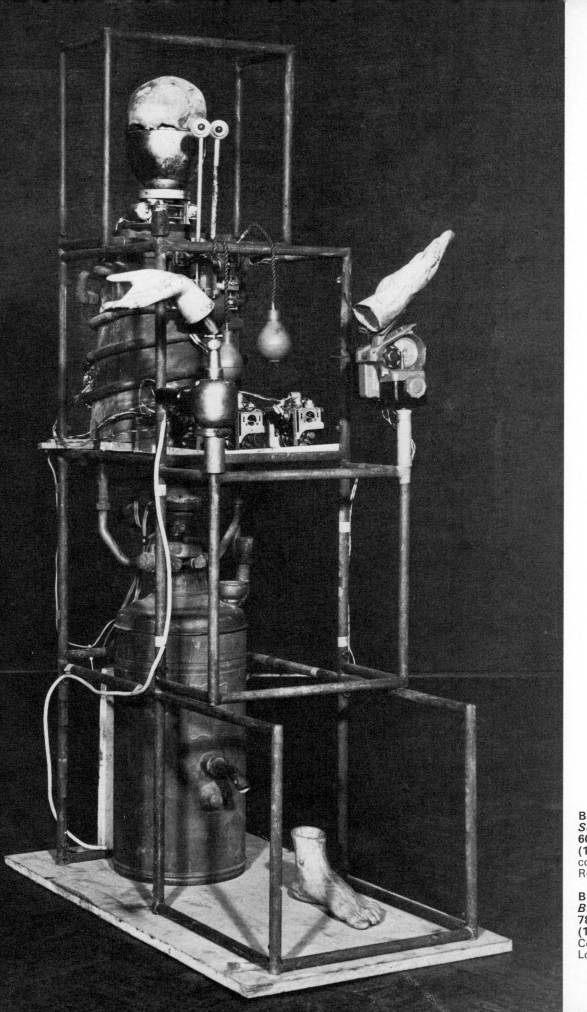

Bruce Lacey:
Superman 2963, 1963,
60×24×36 in.
(152.5×61×91.5 cm.)
collection: Jasia
Reichardt, London

Bruce Lacey: *Boy, Oh
Boy, am I Living!*, 1964,
78×60×12 in.
(198×152.5×30.5 cm.)
Collection: Tate Gallery,
London

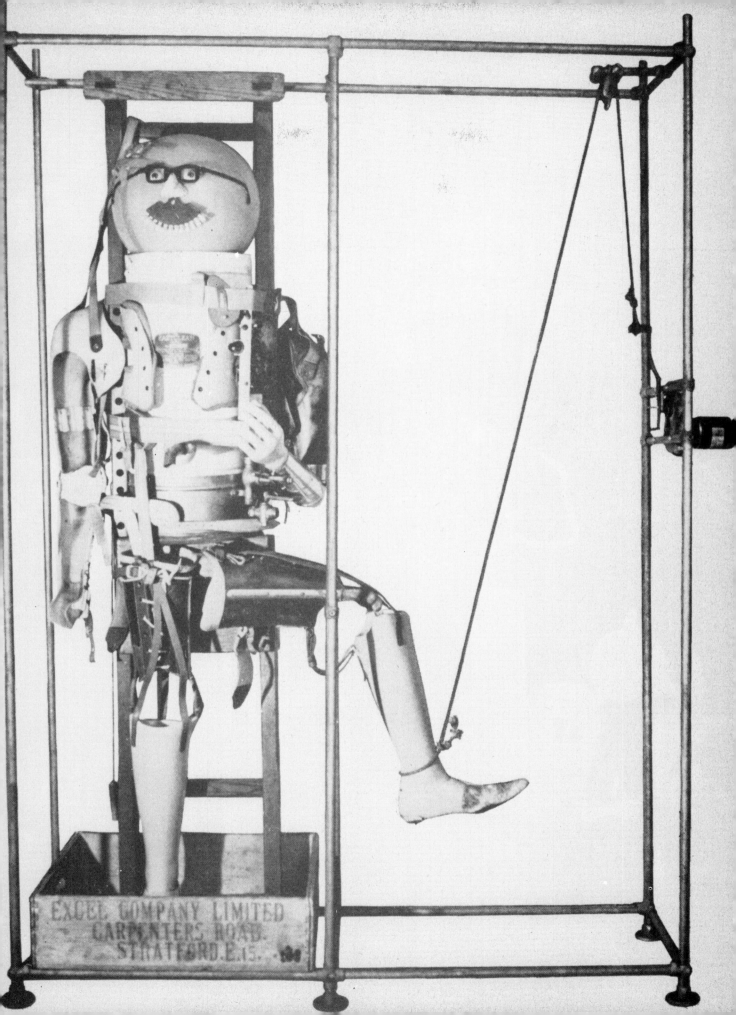

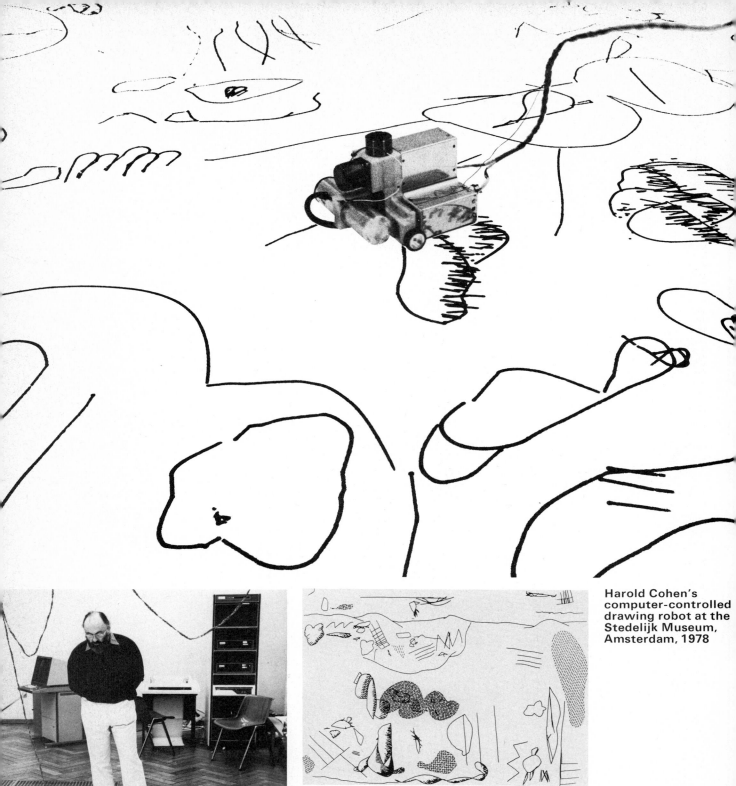

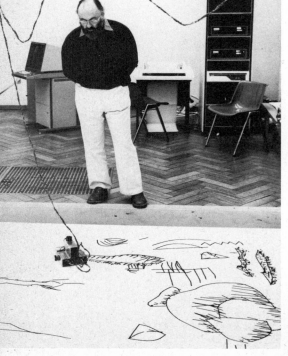

Harold Cohen's computer-controlled drawing robot at the Stedelijk Museum, Amsterdam, 1978

Edward Ihnatowicz. *The Senster*, 1970, 9 ft high with upward reach of 15 ft. The photograph was taken by the artist at the Department of Mechanical Engineering at University College where *The Senster* was put together before its departure for the Evoluon, in Eindhoven, Holland.

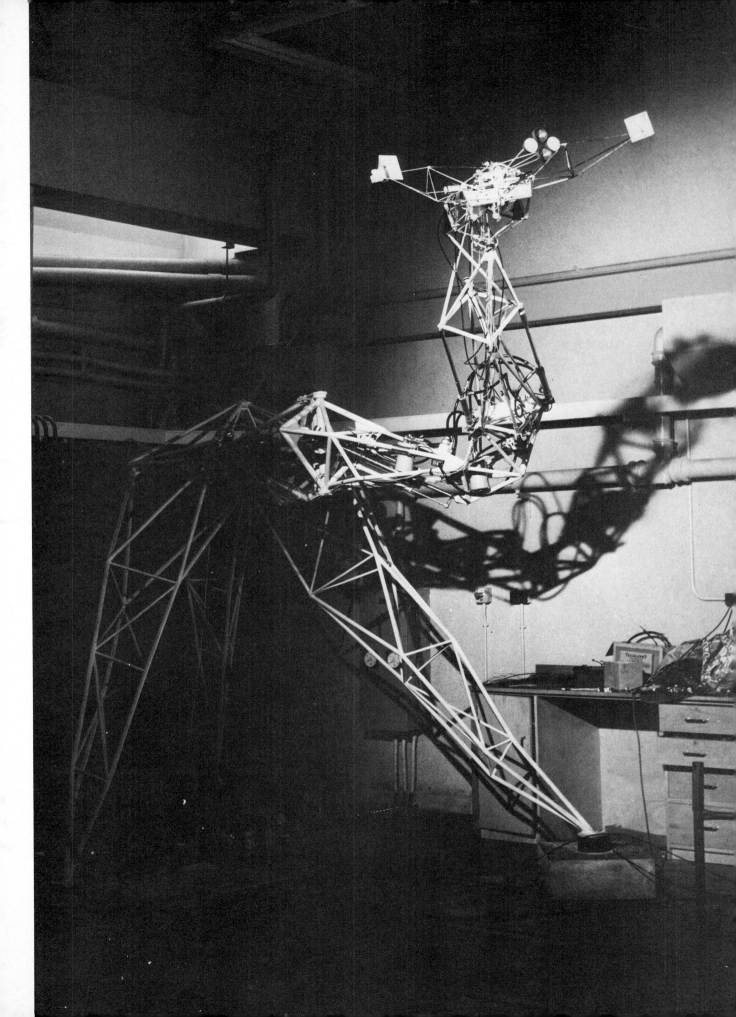

Robots on celluloid

Three of the earliest and most remarkable films about artificially created life originated in Germany: *Homunkulus* (1916), *Metropolis* (1926), and *Alraune* (1918). Between them, they represent themes important enough to have been echoed repeatedly in subsequent films featuring robots.

1. That the creation of artificial life is bound to have devastating consequences.
2. That robots have been made and used by humans with essentially selfish, if not base, motives.

Each of these three films, as well as the films about Frankenstein, involves a scientist/magician, a Faustian figure, who encroaches on a field of activity fraught with danger and doomed from the start: the making of artificial men. In each case the story is about nothing less than life and death, greed, rebellion, love, and madness.

Homunkulus is an artificial man who cannot come to terms with the fact that he does not possess a soul. A creature of pure reason and considerable intellectual powers, he is driven to tyranny through his realisation of his own inhumanity. Having become the dictator of a nation, his plans for conquering the world are brought to a halt when he is killed by lightning. Made with electricity, he is finally destroyed by it. This is the first film to deal explicitly with the dangers of which science and scientists are the instruments.

'I have created a machine in the image of man, that never tires or makes a mistake. . . . Now we have no further use for living workers', boasts Rotwang, the scientist/magician in *Metropolis*. He proceeds to make the double of Maria, the mediator in the underground city where workers toil in indescribable conditions to keep the metropolis above functioning. The purpose of making the 'double' is to counteract the original Maria's efforts to mediate. The robot, when she is first created, looks like a female warrior in gleaming armour. Later, when she is covered in flesh she becomes the indistinguishable copy of Maria and incites workers to violence to destroy the underground city and thereby themselves. When the robot Maria is eventually burnt at the stake by the workers, they see her change in the flames from a woman into a metal robot. Meanwhile, the real Maria escapes and resumes her helpful role.

Alraune, the heroine of *Alraune*, was created by a scientist experimenting with artificial insemination. The daughter of a hanged criminal and a prostitute, she is a great seductress who ruins all

who are in love with her and at the end destroys herself.

Frankenstein's monster (see p. 47) also belongs to this group of films, although it is more often associated with horror movies than science fiction. The monster is made from dead bodies. Accidentally the brain of a madman has been transferred into its skull, which causes the artificial man to be the agent of destruction. However, he is far from being all evil, and Boris Karloff's portrayal shows that even demonic creations can possess human emotions and virtues.

In these four films the artificially created life is the central theme, and the story revolves around the fate of this new being. This is not the case, however, with many films which feature robots, where although they are an essential part of the action, they are rarely its focus. In *Undersea Kingdom* (1936), for instance, robots belong to the setting. Discovering the lost kingdom of Atlantis at the bottom of the ocean, the hero finds a civilization complete with death ray, robots, and horse-drawn chariots. The robot in *The Day the Earth Stood Still* (1951), is there to bring his master back to life after he is killed by people on earth who disbelieve his warnings about the dangers of the use of atomic weapons. In *Forbidden Planet* (1956), the robot Robby, which subsequently appears in *The Invisible Boy*, is a sophisticated houseboy and guard to the philologist Morbius and his daughter, who are the sole human survivors on the planet Altair. Robby has no central role but he became so popular that he was subsequently used in other films. In his appeal, he was compared to the dog Lassie, something that has also been said about such robots as Threepio and Artoodetoo (see below).

Other robots which made their mark include Gog, who had five arms and performed menial tasks in an underground laboratory; and three squat robots in *Silent Running* (1971), two of whom become the card-playing companions of the last remaining human, who entrusts one of them to look after the greenhouse, representing the surviving plant life which is sent out of the earth's orbit for its own safety. In *Star Wars* (1977), the two robots, like a well-attuned comedy team, become the focus of the film although they are only the unwitting participants of an intergalactic drama (see colour p. 24). Threepio, the fussy, nervous robot, whose appearance is based on the metallic robot of *Metropolis*, and the cylindrical Artoodetoo, are trying to survive in a situation which they hardly understand. The success of the film rests to an enormous extent on the robots, which have the appeal and charm of pets. Despite the importance of the tasks which they perform,

In *Metropolis* the scientist/magician Rotwang (Rudolf Klein-Rogge) looking at Maria (Brigitte Helm) encased in a plastic cylinder, whose likeness is being transferred to the robot in the background

Threepio and Artoodetoo, like Robby, remain beautiful, large, complicated and desirable toys. *Star Wars* does not touch upon the subject of the relationship of men and robots or on the plight of artificially created people.

Two films which have something significant to say about both men and intelligent machines are *Westworld* (1973) and *Demon Seed* (1977). The first is about a Disneyland for adults, where people can go to live out their fantasies, offering its customers murder, violence, and sex. The robots, indistinguishable from human beings, engage in realistic gun battles against the human visitors, who dress up as cowboys. The robots are always defeated and go for repair each night. The principal robot gunslinger (Yul Brynner) gets out of control, pursues the hero and the film ends with Westworld reduced to a wasteland. The film is concerned with the idea that even licensed misdeeds have their moral consequences, and that nobody may decree what anyone may be allowed to do to anyone, or to anything, else.

In *Demon Seed*, a massive computer complex also manifests human attributes. Having surpassed by far the potential of the human brain, it wants to achieve immortality, but it cannot be immortal since it is not subject to death, and apparently it is only by conquering death that immortality can be achieved. With the aid of a robot consisting of an arm attached to a wheelchair, the electronic brain uses the resources of a laboratory to impregnate the wife of the scientist who built him. A child is born twenty-eight days later, but when it is taken out of its protective covering, it is about four years old. 'I am alive', says the baby girl in a deep baritone voice; the one thing her 'father' was not. By now, humans have begun to fear him, his existence is threatened, and, as his aim has already been achieved, he switches himself off.

The most successful robot films are those which touch on the discrepancy between what we feel (empathy with the machine) and what we know (that it is only a machine). It is notable that when watching the film the gap between one and the other gradually narrows and we experience the same emotions whether what we see is the man or the machine. We start to differentiate only as a result of subsequent rationalization.

Robots in films – some landmarks

1897 Gugusse and the Automaton by Georges Méliès
1910 Frankenstein by J. Searle Dawley. Edison
1914 Der Golem by Paul Wegener and Henrik Galeen. Bioskop
1916 Homunkulus (six-part serial based on the novel by Robert Reinert) by Otto Rippert. Bioskop
1918 Alraune (first of several film versions of the book by Hans Heins Ewers). Neutral Film

1920 Der Golem (based on the book by Gustav Meyrink) by Paul Wegener and Karl Boese. UFA
1926 Metropolis (based on the book by Thea Harbou) by Fritz Lang. UFA
1931 Frankenstein (adaptation of Mary Shelley's novel by Robert Florey, with Boris Karloff) by James Whale. Universal
1935 Bride of Frankenstein (with Karloff) by James Whale. Universal
1935 Phantom Empire (serial) by Otto Brower and B. Reeves Eason. Mascot
1936 Undersea Kingdom (serial) by B. Reeves Eason and Joseph Kane. Republic
1939 Son of Frankenstein (with Karloff) by Rowland van Lee. Universal. (More than forty films about the Frankenstein monster have been made to date.)
1944 Dreams that Money Can Buy by Hans Richter (includes sequence by Fernand Léger about a girl with a manufactured heart). Art of This Century Films
1951 The Day the Earth Stood Still (based on long story 'Farewell to the Master' by Harry Bates) by Robert Wise. Fox
1951 Captain Video (serial) by Spencer Bennet and Wallace A. Grissell. Columbia
1952 Zombies of the Stratosphere (serial) by Fred C. Brannon. Republic
1954 Tobor the Great by Lee Sholem. Republic
1954 Gog (based on story by Ivan Tors) by Herbert L. Strock. Ivan Tors
1954 Target Earth! (based on story 'The Deadly City' by Paul W. Fairman) by Sherman Rose. Allied Artists
1956 Forbidden Planet (with plot loosely based on *The Tempest*) by Fred McLeod Wilcox. MGM
1957 The Invisible Boy (based on a story by Edmund Cooper) by Hermann Hoffmann. Pan
1968 2001 : A Space Odyssey (based on 'The Sentinel' by Arthur C. Clarke) by Stanley Kubrick. MGM
1970 THX 1138 by George Lucas. Zeotrope Productions
1971 Silent Running by Douglas Trumbull. Universal
1973 Westworld by Michael Crichton. MGM
1974 Terminal Man (based on the novel by Michael Crichton) by Michael Hodges. Warner Bros
1976 Logan's Run by Michael Anderson. MGM
1977 Demon Seed (based on the novel by Dean Koontz) by Donald Cammell. MGM
1977 Star Wars by George Lucas. Lucasfilm

Boris Karloff in *Son of Frankenstein*, 1939, the last of the three Frankenstein films in which he appeared. Karloff created the prototype of the monster as a creature who is wicked because it is profoundly unhappy.

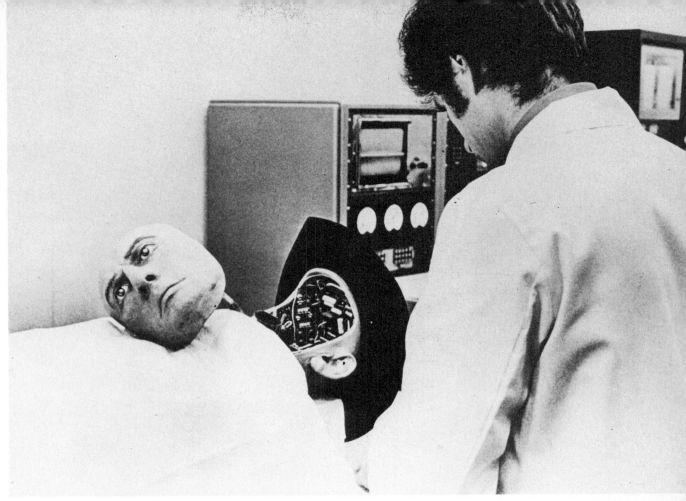

Robots are mended at night in the laboratory in Westworld.

Scene from *The Evil of Frankenstein*, 1963, in which the monster is about to be brought to life with the aid of electric lightning. The equipment is very similar to that in *Metropolis*.

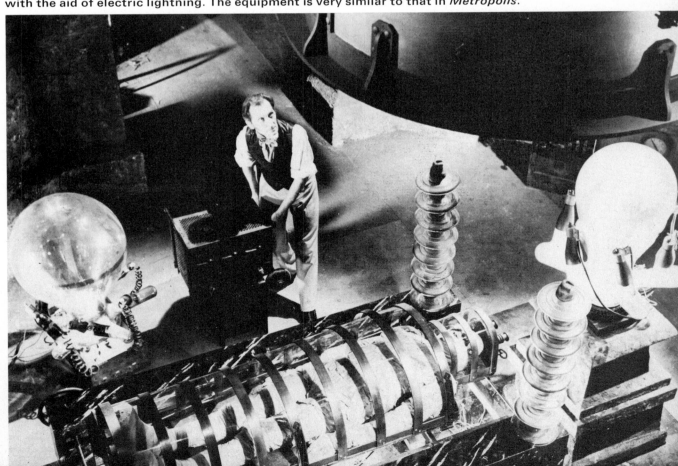

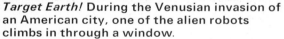

Target Earth! During the Venusian invasion of an American city, one of the alien robots climbs in through a window.

Captain Video. The same robot as in *Phantom Empire* participates in an adventure serial ▲ ▲ about a dictator of another planet who wants to rule the universe but is finally outwitted.

Phantom Empire is a cowboy film with robots, ▲ television sets, death rays and supermen, who inhabit the subterranean city of Murania.

Zombies of the Stratosphere. The robot, ▶ created by one of the Zombies, is attacking a member of the Inter-Planetary Patrol.

Gog. One of the robots which eventually goes berserk in an underground atomic laboratory

Undersea Kingdom is the lost city of Atlantis, in which period costumes and the latest technology coexist.

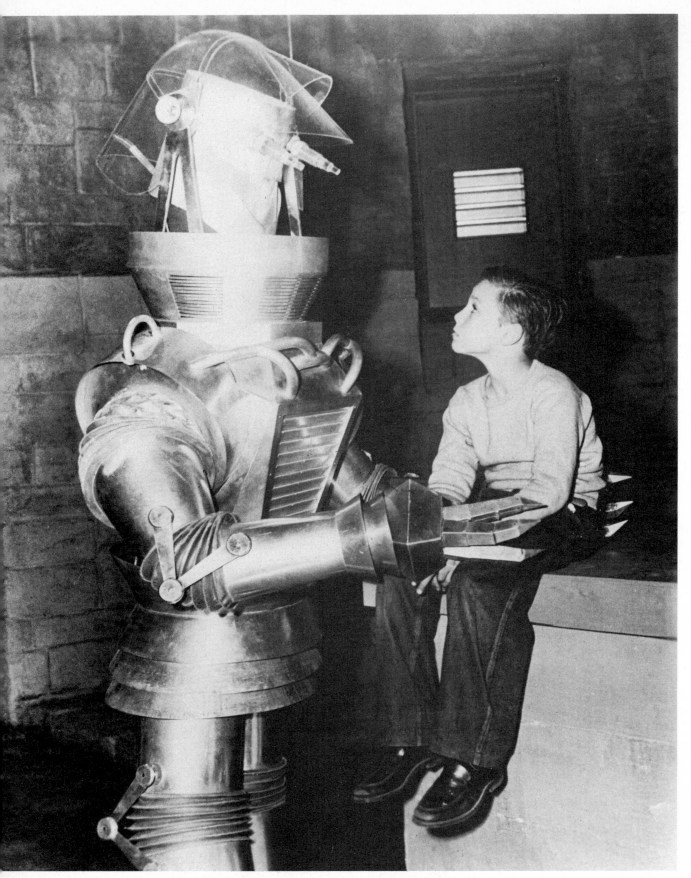

Tobor the Great, programmed with special concern for children, is the first film robot to have been built specifically for space missions.

Robby the Robot appears in *Forbidden Planet*, as well as in *The Invisible Boy*, in which he participates in the adventures of a scientist and his small son. ▶

Murder Mystery 1

The following *Murder Mystery 1*, consisting of 2100 words, was produced in nineteen seconds by UNIVAC 1108, and first presented at the International Conference on Computers in the Humanities, Minneapolis, July 1973.

The authors of *Murder Mystery 1*, or rather of the program used, i.e. the Novel Writer Simulation Program, are: Sheldon Klein, J.F. Aeschlimann, D.F. Balsiger, S.L. Converse, C. Court, M. Foster, R. Lao, J.D. Oakley, and J. Smith, of the Computer Sciences Department and Linguistics Department, University of Wisconsin, Madison.

The program allows the plot and the behaviour of the individual characters to be determined to an extent at random as the story progresses.

With the exception of the addition of capital letters and increased spacing between paragraphs, *Murder Mystery 1* has not been altered in any way.

Wonderful smart Lady Buxley was rich. Ugly oversexed Lady Buxley was single. John was Lady Buxley's nephew. Impoverished irritable John was evil. Handsome oversexed John Buxley was single. John hated Edward. John Buxley hated Dr Bartholomew Hume. Brilliant Hume was evil. Hume was oversexed. Handsome Dr Bartholomew Hume was single. Kind easy going Edward was rich. Oversexed Lord Edward was ugly. Lord Edward was married to Lady Jane. Edward liked Lady Jane. Edward was not jealous. Lord Edward disliked John. Pretty jealous Jane liked Lord Edward.

Well to do Ronald was kind. Lusty Ronald was married to Cathy. Handsome Ronald loved Catherine. Ronald liked Hume. Ronald disliked James. Easy going lusty Cathy was kind. Beautiful jealous Catherine loved Ronald. James was Ronald's partner. James hated Ronald. Evil violent James was dumb. Impotent ugly James was married to Marion. Well to do jealous James disliked Marion. James disliked Dr Bartholomew Hume. Unpleasant violent Marion was smart. Beautiful Marion was impoverished. Jealous oversexed Marion hated James. Marion disliked Florence.

Florence was Lady Buxley's companion. Wonderful Florence was easy going. Beautiful oversexed Florence was single. The smart unpleasant butler was lusty. Poor brave butler was single. The dumb maid was good. Pretty poor Heather was single. Ugly violent cook was single. The cook was poor.

The day was Tuesday. The weather was rainy. Marion was in the park. Dr Bartholomew Hume ran into Marion. Hume talked with Marion. Marion flirted with Hume. Hume invited Marion. Dr Hume liked Marion. Marion liked Dr Bartholomew Hume. Marion was with Dr Bartholomew Hume in the hotel. Marion was near Hume. Dr Hume caressed Marion with passion. Hume was Marion's lover. Lady Jane following them saw the affair. Jane blackmailed Marion. Marion was impoverished. Jane was rich.

Marion phoned Jane in the morning. Marion invited Jane to go to a theater. Jane agreed. Jane got dressed for the evening. They met them in the theater. Jane introduced Lord Edward during an intermission to Marion.

The day was Wednesday. The weather was windy. Lady Jane was in the tennis court. John ran into Lady Jane. John talked with Jane. Lady Jane flirted with John Buxley. John Buxley invited Lady Jane. John liked Lady Jane. Lady Jane liked John. John Buxley was with Jane in a movie. John was near Lady Jane. Jane caressed John Buxley with passion. Lady Jane was John's lover. Cathy following them saw the affair. Cathy blackmailed Lady Jane. Jane was well to do. Lady Catherine was rich.

Lady Catherine invited Jane to play bridge. Lady Catherine told Marion to come with Lady Buxley. Jane asked them to sit down. Lady Jane brought the cards. Jane offered drinks. Lady Buxley asked for whiskey on the rocks. The others had coffee with cookies. Jane shuffled the cards. Lady Jane started a game. Marion casually signaled Lady Buxley with hands. Jane noticed it. Lady Jane suspected that they cheated. Jane watched them closely. Marion won the game with Lady Buxley. Jane was upset with Catherine. Lady Jane disliked Marion.

The day was Thursday. The weather was rainy. A small pub was on a corner. John Buxley was in the pub. John Buxley asked for whiskey on the rocks. John got a drink from the barman. John talked with Hume near the bar. Hume sang the Beatles's song. John Buxley was drunk. James said that Marion commited adultery. Hume thought that James was drunk. James was depressed. James left the pub. Edward said that Lady Jane commited adultery. John Buxley thought Lord Edward was drunk. Lord Edward was depressed. Lord Edward left the pub.

The day was Friday.

Lady Buxley had a big house. Lady Buxley's house had a pretty fragrant garden. A green house was in the garden. The garden was near the tennis court. The house had a big bright dining room. The house also had a pleasant parlor. A cool dark musty library was near the parlor. The time was evening. Lady Buxley gave a party. The party lasted for a weekend.

Lady Buxley talked with Florence.

Marion arrived with James.

Catherine arrived with Ronald.

Edward arrived with Jane.

Dr Hume arrived. Dr Bartholomew Hume joined a conversation.

Catherine talked with Dr Bartholomew Hume. Dr Bartholomew Hume flirted with Lady Catherine. Dr Bartholomew Hume said that Lady Catherine was beautiful. Dr Hume wanted to seduce Catherine. Hume told a joke. Catherine laughed.

Lady Buxley talked with Ronald. Florence talked with Dr Bartholomew Hume. Dr Hume flirted with Florence. Dr Bartholomew Hume flattered Florence. Florence was very aroused. Dr Bartholomew Hume liked Florence. Florence liked Hume.

The servants went to bed.

John Buxley arrived. Lady Buxley greeted John Buxley. John joined the conversation.

John Buxley talked with Jane. John Buxley casually mentioned politics. Lady Jane discussed politics with John Buxley. Lady Jane said that the weather was nice.

Lord Edward talked with Lady Jane. Florence talked with Edward. Edward flirted with Florence. Lord Edward wanted to seduce Florence. Lord Edward smiled at Florence. Florence smiled at Lord Edward. Jane saw that Edward whispered to Florence. Lady Jane was angry. Lord Edward saw that Lady Jane was angry.

Marion talked with Lord Edward. Lord Edward flirted with Marion. Lord Edward said that Marion was beautiful. Lord Edward smiled at Marion. Edward gently touched Marion. Lord Edward whispered to Marion. Edward liked Marion. Marion liked Edward. James saw that Marion talked with Edward. Jane saw that Edward whispered to Marion. Jane was angry. Jane saw that Edward smiled at Marion.

Everyone went to bed.

The day was Saturday. The sun rose. The servants got up. The cook went to the kitchen. The cook prepared a breakfast. Clive followed the cook. Clive seduced Maggie in the kitchen.

The day was beautiful. They got up. They got dressed. They went down to the breakfast.

Florence talked with Ronald. Ronald said that Florence looked well. Florence casually mentioned business. Ronald hated conversations about business.

The breakfast was over. James talked with Lady Buxley. James casually mentioned a music. Lady Buxley discussed the music with James.

Everyone went to the parlor.
James talked with Dr Hume. Hume argued with James. James said that Hume was idiotic. Hume threatened to hit James. Dr Bartholomew Hume cursed James. James hit Dr Bartholomew Hume in the nose. Dr Bartholomew Hume tried to grab James. James pushed Hume. Hume threatened to kill James. Dr Bartholomew Hume hit James. James hated Dr Hume.

Dr Hume asked Lord Edward to play chess. Edward agreed. Lord Edward went to the study with Dr Hume. They played chess. Hume was a good player. Lord Edward played chess well.

Florence talked with John. John flirted with Florence. John wanted to screw Florence. Florence smiled at John Buxley.

James talked with John. John laughed. John Buxley said that James looked well.

Ronald talked with James. James argued with Ronald. Ronald said that James was idiotic. James threatened to hit Ronald. Ronald hit James. James kicked Ronald in the belly. Ronald groaned softly. Ronald hit James in the nose. James tried to grab Ronald. Ronald pushed James. Ronald struggled with James. James threatened to kill Ronald. James hit Ronald. Ronald hated James.

Lady Buxley talked with Florence.

The cook went to the kitchen. Maggie prepared lunch.

Ronald talked with Lady Buxley.

Clive announced lunch. Edward stopped playing chess. Dr Bartholomew Hume stopped playing chess.

Everyone went to the dining room. Everyone sat down. Clive served the food. Lunch started.

Florence talked with Hume. Florence casually mentioned fashion. Dr Bartholomew Hume hated the conversations about fashion.

Lunch was over. The men went to the parlor. The men smoked cigars. The women went to the drawing room. The women drank whiskey.

Everyone went to the parlor. Marion decided to go for a walk. Marion smiled at Edward. Edward saw that Marion went to the garden. Edward followed Marion. Jane saw that Edward followed Marion. Jane thought that Lord Edward loved Marion. Jane followed Lord Edward. Lord Edward met Marion.

Edward kissed Marion. Marion caressed Edward. They went to the green house. Lady Jane followed them. Marion undressed. Edward screwed Marion. Edward commited adultery. Marion commited adultery. Lady Jane was enraged. Jane entered the green house. Jane yelled at Lord Edward. Jane cried. Jane threatened to kill Lord Edward. Marion was embarassed. Lord Edward asked Lady Jane to forgive Lord Edward. Everyone went to the house.

Marion talked with John Buxley. John Buxley flirted with Marion. John Buxley gently touched Marion. Marion smiled at John. John Buxley wanted to seduce Marion. Marion wanted to seduce John Buxley. James saw that Marion talked with John. James was mad at John. James overhearing Marion was angry. Marion saw that James was upset. Marion talked with James.

The butler announced tea.

Everyone went to the garden. The butler served tea. The day was cool. The sky was cloudy. The garden was nice. The flowers were pretty. Marion complimented Lady Buxley.

Ronald talked with Marion.

Tea time was over.

Everyone went to the parlor.

The cook went to the kitchen. Maggie prepared dinner.

Dr Hume asked Edward to play tennis. Edward agreed. Lord Edward went to the tennis court with Dr Hume. They played tennis. Dr Hume was the good player. Edward played tennis well.

The butler announced dinner.

Dr Bartholomew Hume stopped playing tennis. Edward stopped playing tennis.

Everyone went to the dining room. Everyone sat down. The butler served the food. Supper started.

Marion talked with Florence. Florence argued with Marion. Marion said that Florence was idiotic.

Florence talked with Lady Buxley.

Supper was over. The men went to the parlor. The men smoked fat smelly stogies. The men drank sherry. The women went to the drawing room. The women gossiping drank coffee.

Everyone went to the parlor.

Marion talked with Jane.

James went to the library. James read the good paperback. Edward asked Ronald to play tennis. Ronald agreed. Ronald went to the tennis court with Lord Edward. They played tennis.

John suggested the game of bridge. Lady Buxley agreed. Dr Bartholomew Hume agreed. Jane agreed. They played bridge.

The servants went to bed. Everyone went to bed.

James stopped reading the book.

Ronald beat Lord Edward at tennis. Lord Edward stopped playing tennis. Ronald stopped playing tennis.

John Buxley cheated at bridge.

John cheated at bridge.

The card game was over.

John awoke. John Buxley got up. John planned to meet Marion. John entered the corridor. Marion got up. Marion went to the hall. James knew the plan. James decided to follow them.

John Buxley kissed Marion. Marion kissed John. They went to the library. James followed them. Marion undressed. John Buxley screwed Marion. Marion commited adultery. James was enraged. James entered the library. James yelled at John. James threatened to kill John Buxley. Marion was embarassed. Marion cried. Everyone went to bed.

James was very rich. Clive was impoverished. Clive wanted the money. The butler was related to James. The butler decided to poison James. Clive thought that Clive inherited the money. Clive knew that James drank a milk. Clive poisoned the milk. James drank the milk. James went to bed. James died. The others thought that James was asleep. Clive removed the fingerprints. The butler returned the bottle.

Ronald awakened. Ronald got up. Ronald thought that the day was beautiful. Ronald found James. Ronald saw that James was dead. Ronald yelled. The others awakened. The others ran to Ronald. The others saw James. Everyone talked. Heather called the policemen. Hume examined the body. Dr Bartholomew Hume said that James was killed by poison.

John talked with Edward about the murder.

Edward talked with Maggie about the murder. Maggie was upset about the murder.

The cops arrived. The cops were idiotic. A detective examined the corpse. The policemen looked for hints in the bathroom. Dr Bartholomew Hume also looked. Edward tried to calm Marion.

The policemen questioned Dr Bartholomew Hume. The detective asked questions. The policemen searched the garden. The policemen tried to find clues. Marion cried.

Dr Bartholomew Hume searched stairs. Hume looked for hints. Dr Hume questioned Lady Buxley. Dr Hume knew that Lady Buxley told the truth. Florence talked with Heather about the murder. Marion cried.

The policemen questioned Ronald. The inspector suspected Ronald. The inspector asked the stupid questions. The policemen searched the parlor. The policemen tried to find hints. Florence was upset.

Dr Bartholomew Hume searched the dining room. Dr Bartholomew Hume looked for hints.

The cops questioned Heather. The detective asked the stupid questions. Dr Hume questioned Heather. Dr Hume knew that Heather told the truth. The cops searched the tennis court. Clive talked with Ronald about the murder. The butler said that James was kind. The cook talked about the murder.

Dr Bartholomew Hume searched the bathroom. Dr Hume looked for clues. Marion cried.

Dr Hume questioned Florence. Hume knew that Florence told the truth. Dr Bartholomew Hume got information from Florence. The cops searched the bathroom. The cops found a thread. The thread was a misleading clue. Lady Buxley talked with John about the murder. Lady Buxley said that James was kind. Dr Hume was upset.

Dr Bartholomew Hume searched the library. The cops questioned John Buxley. The detective asked the stupid questions. Hume questioned the cook. Dr Bartholomew Hume knew that Maggie told the truth. Hume got information from the cook.

Hume went to the bathroom. Dr Hume found the bottle. Hume knew the murderer. Hume asked everyone to go to the parlor. Dr Bartholomew Hume said that the murderer was in the room. Everyone was surprised. Everyone talked. Dr Bartholomew Hume said that James was killed by poison. Hume said that the butler killed James. Everyone was shocked. The butler drew a pistol. Clive headed for the door. Dr Bartholomew Hume followed Clive. The butler shot at Hume. Dr Bartholomew Hume grabbed a paperweight. Dr Bartholomew Hume threw the paperweight at Clive. The paperweight hit Clive in the head. Clive fell. Dr Bartholomew Hume took the gun. The policemen took Clive. Ronald congratulated Hume. Clever Dr Hume solved the crime.

Drawing by Hans Küchler

Performing robots

The Osaka Demonstration Robot consists of a head, body, base and two arms of different lengths. The head contains two control rooms. In the first ambient data is collected and processed, and then transmitted to the main control room from which the robot receives instructions to respond by emitting smoke, smells, light, and sounds. The body of the robot can rise up to a height of twenty-four feet. When the body is up, the base becomes a stage and it is then possible for the body of the robot to go through a repertoire of movements.

1 Main control room
2 TV camera
3 Head
4 Hatch
5 Data collection control room
6 Neck
7 Microphone collecting ambient sounds
8 Speaker
9 9-metre arm with flexible extension and contraction movement
10 Lifting hook
11 Observation window
12 Strobe
13 Main body
14 Base
15 Footlights
16 Bumper
17 Supporting pillars for lifting main body up and down
18 Performance deck
19 and 22 Nozzles for water spray
20 Illumination lamps
21 Window for spot light
23 Roof deck
24 Performance arm with a platform for carrying people
25 The platform, otherwise called 'finger deck'
26 Base wheels

Diagram of the performing robot at Expo '70 in Osaka

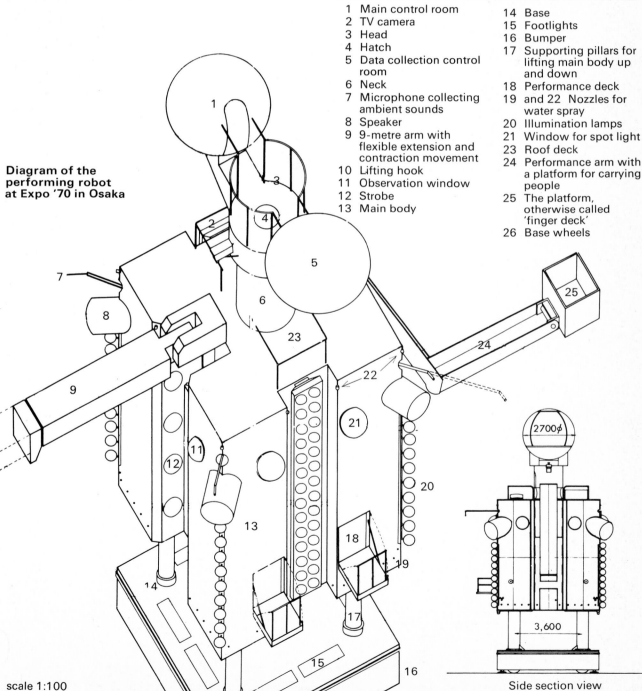

scale 1:100

2700φ

3,600

Side section view

During the fourteenth century the counts of Artois appointed a 'Master of the Amusement Machines and of the Painters' for their castle at Hesdin, and constructed a gallery of machines for the amusement of their guests, and especially the members of the household. Most of the machines sprayed those looking at them with water, flour or soot. In his book *The Rise of the Artist*, Andrew Martindale reports: '. . . those entering were spoken to by a wooden figure of a hermit by the entrance. Above the ceiling was a machine for producing water like rain from the sky, and thunder, snow and lightning "such as is seen in the sky"; and in the floor at one point was a trap down which those escaping from the "rain" fell, straight into a large sack where they were smothered in feathers. Various carved figures throughout the gallery projected water without warning at onlookers. Three jets near the entrance covered people who stopped in front of them with flour. There was an imitation window which, on being opened, revealed a figure who soaked the onlookers with water and closed the window again.'[1]

If any royal household, bank or museum were to decide to have a gallery of twentieth-century amusement machines, it is unlikely that any of the mechanical devices would indulge in slapstick. On the contrary they would provide a somewhat refined entertainment of songs, dances and witticisms for which one would presumably feed their coin-operated bellies. If a visitor to such a gallery wondered why twentieth-century technology was so exceedingly primitive, it would no doubt be explained that entertainment robots were not thought of as a high priority, and refer the visitor to achievements in robotics in industry, medicine, and space research.

The heyday of the performing robots was in the 1930s although they were to be seen at many international fairs before and since. The London Radio Exhibition of 1932 presented several robots which could bow, make speeches, sing, smoke cigars and read newspapers. The chromium-plated Alpha, which with Eric and Elektro is one of the best-known performing robots, was made by Harry May for the Mullard Valve Company. It could tell the time in several languages and read aloud the daily newspapers, which were prerecorded on a set of records early each morning. Alpha had a particularly powerful voice which, the visitors to the exhibition were assured, could shatter all the windows in Olympia.

Eric (on the cover), the robot inspired by *R.U.R.* and made by Captain W. H. Richards in 1928, opened the annual exhibition of the Model Engineer Society. On that occasion, when asked to address the audience, he rose slowly to the accompaniment of whirring noise, bowed stiffly from the waist, his eyes flashed, a spark jumped across his teeth, and turning his head from side to side, Eric began his speech. The words were relayed to the loudspeaker in his throat via a wireless from a small broadcasting plant offstage. His working mechanism consisted of batteries, two electric motors, and a system of belts, shafts and pulleys.

Elektro, one of the most photographed mechanical men, was produced by Westinghouse Electric Corporation in 1939 for the New York World's Fair. Constructed from aluminium on a steel frame, Elektro was capable of performing twenty-six movements and responded to commands spoken into a microphone. Each word set up vibrations which were converted into electrical impulses; which in turn operated the relays controlling eleven motors. A series of words properly spaced selected the movement Elektro was to make. Two-word commands started an action. One-word commands stopped it. Four words returned all relays to their normal positions. His fingers, arms and turntable for talking were operated by nine motors and another small motor worked the bellows so that the giant could smoke. The eleventh motor drove the four rubber rollers under each foot, enabling him to walk. Elektro's faithful companion, the dog Sparko, with two motors, could beg, bark and wag his tail.

The performances, however, were not always confined to the stage. A French journalist visiting London in 1906 reports in *La Nature* having seen a life-size male doll walking in Piccadilly. Elegantly dressed, with blond curls and staring eyes, the strange creature was accompanied by three attendants, followed by several policemen and a vast crowd. One of the attendants was seen to put his hand under the artificial man's jacket, wind something up, causing it to change direction, sit down on a chair provided by a second attendant, or move his arms. Occasionally one of the limbs or the head of the automaton would be removed, at which point it would stop and when the part was replaced, the promenade would be resumed. During the following half-hour traffic came to a standstill. Enigmarelle, who we learn was about to appear at the Hippodrome, was unexpectedly addressed by a policeman who had been looking ill at ease for some time. The Frenchman ends his article with the report that '. . . on the following day London had the pleasure to

[1] Andrew Martindale. *The Rise of the Artist in the Middle Ages and Early Renaissance*. Thames & Hudson, London, 1972

Alpha at the London
Radio Exhibition in
1932

ONOFF the Wonder
Robot of the Wonders
of the World Museum
in California

learn that for the first time in the history of the world, an automaton was summoned before a police tribunal on the charge of causing a disturbance on the public highway'.

Even stranger stories are told by Rolf Strehl.[2] At the Chicago World's Fair, Roland Schaffer exhibited an artificial man which sawed wood, hammered nails and carried trunks from one place to another. One day, while sitting at his desk looking through some drawings he heard a noise and turned round to see the robot marching straight for him, swinging an iron club normally used for forging. Within seconds the robot killed Schaffer, demolished the entire laboratory, and collapsed. The robot was discovered by a showman who later resurrected it. According to the electro-mechanical workshop where the robot was reconstructed, it had inside it a mechanism which controlled its balance and was powered by electricity and compressed air. The head contained an aerial which received wireless signals from a special transmitter controlling its movements, suggesting that an unidentified human was responsible for the murder.

There is another story of a robot involved with its maker's death. In 1946 a Milwaukee engineer was adjusting the arm of a robot which contained more than two hundred electronic valves, when the complicated mechanism failed and the robot crushed him under its great weight.

The systems of the robots described so far were very much simplified in 1939 by Mechanical Man Inc. who started to manufacture automaton salesmen. These small dolls were usually powered by tiny electric motors, with the movement of the different parts of the body achieved by rod linkages between discs driven by gears or pulleys, and a pump with a crank-operated piston supplying the suction through a small tube, for puffing a cigarette. Miniature mannequins paraded up and down department store windows dressed in the latest fashions, and a bowing figure made of oil tins and an oil drum used to greet motorists stopping at garages. A sports shop, meanwhile, displayed a four-foot hunter who held his gun at rest, then, as the pheasant rose, he turned and tilted his head forward, aimed and fired, and as a red light flashed the bird disappeared.

The making of robots as a pastime continues. Two books, one by A.H. Bruinsma[3] and the other by David L. Heiserman,[4] provide useful guidelines. Bruinsma deals with two sorts of robots: one resembling an animal, and the other suitable for playing games of noughts and crosses. Heiserman gives instructions for constructing

Buster, which according to one of his readers would take between 1000 and 1500 hours to make, although the robot would be able to perform a certain number of tasks before completion. The author says about Buster: 'He is more like an animal than a machine, he has some basic reflex mechanisms, a will of his own, and a personality of sorts.' The robot can be fully operational without human intervention and can set its own goals. Also, 'it can interact with a human operator, provided he doesn't have any other needs that are more urgent at the time'.

Buster comes in three degrees of sophistication. The most complex version has tracking functions. An alarm will sound the moment the battery reaches the first alarm level and the robot will begin to search for a battery charger. It will follow any light surface beneath its frame. A white tape on the floor will provide a sufficient stimulus for Buster operating in tracking mode to deliver a cup of coffee to the right place. The tag-along function will make him follow anyone who shines a light on his phototransistor eyes. A specific sound could produce a similar effect. With a two-way wireless-microphone, Buster can strike up a conversation with someone at a considerable distance which can then be relayed, or carry a battery-powered television camera, while an operator can manoeuvre him over long hallways and around obstacles.

In recent years two very different performing robots have been in the news. One is the gigantic robot at the Festival Plaza of the Expo '70 in Osaka, which moved beside a pool of water, flashing lights, rotating its head and gesticulating. It was big enough to hold several people in the palm of its hand, or rather on the platform which substituted for a hand. Unlike most other robots at international exhibitions, it was designed as part of the architectural setting and became an integral element of a cybernetic environment devised by Arata Isozaki. Here, the robot became like a member of an orchestra, an instrument responding to the music around it and contributing to the overall sound.

The second robot, called ONOFF, is a modest one. Built from scrap by Clayton Bailey of the Wonders of the World Museum at Port Costa, California, it is an effective publicist for the museum. It asks people in the street to insert coins in a slot, in exchange for which it produces and

[2] Rolf Strehl. *The Robots are Among Us*. Arco Publishers, London and New York, 1955
[3] A.H. Bruinsma. *Practical Robot Circuits, electronic sensory organs and nerve systems*. Philips Technical Library, Eindhoven, 1959
[4] David L. Heiserman. *Build Your Own Working Robot*. Tab Books, Blue Ridge Summit, Pennsylvania; and Foulsham-Tab, Slough, 1976

Unidentified robot with a nude

distributes postcards of itself. With suitable sound effects and flashing lights it then invites everyone to follow him, which they usually do, and then heads straight for the museum and its very large collection of toy robots.

With the exception of the Osaka giant, performing robots are examples of a decidedly primitive technology and belong to the fairground and the theatre where illusions are created with the willing participation of the public. In terms of comparison with the past, none is as technically impressive in the context of its time as the Vaucanson duck must have been in 1738 when it was first shown in Paris. But then, there are reasons for this because priorities of robotics lie elsewhere.

Elektro, the robot of Westinghouse Electric Corporation

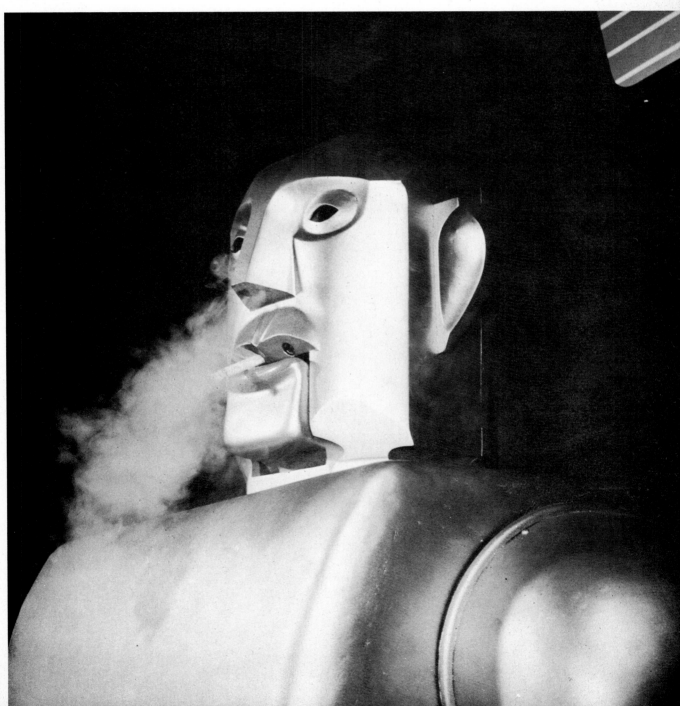

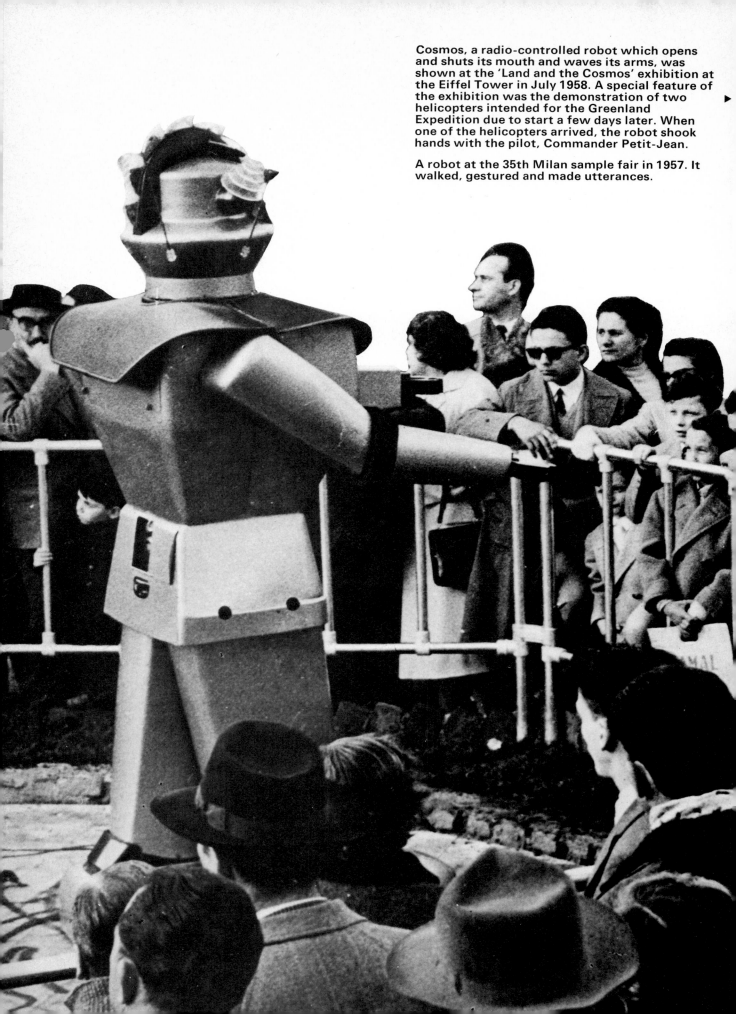

Cosmos, a radio-controlled robot which opens and shuts its mouth and waves its arms, was shown at the 'Land and the Cosmos' exhibition at the Eiffel Tower in July 1958. A special feature of the exhibition was the demonstration of two helicopters intended for the Greenland Expedition due to start a few days later. When one of the helicopters arrived, the robot shook hands with the pilot, Commander Petit-Jean.

A robot at the 35th Milan sample fair in 1957. It walked, gestured and made utterances.

A robot at the international fair in Budapest in 1937

Tentative and metallic love

There is no great love story about a woman and a robot which rings true, which is neither awkward nor crude, and which does not finish before it starts. There are several moving and ingenious stories about men and artificial women, but the other way round there is little to report. Affection of women for robots is hinted at but it is rarely completely outspoken. For instance, when the young programmer Xanthe[1] takes her robot Xanthippus for a walk in the woods and he cannot make sense of the signals of nature, and collides with trees, she is moved by his helplessness in the real world. Clearly her emotions become engaged because she tells herself severely: 'Xanthe, you're becoming sentimental; please remember that a robot is a robot, a mechanical thing.'

Reticence is also the key to one of the most important female characters in robot stories, viz. Asimov's Susan Calvin. Born in 1982, the same year that the US Robot and Mechanical Men, Inc. come into being, she becomes a robopsychologist with the firm in 2008, and stays until her retirement. She feels that with the advent of robots mankind is no longer alone, now we have creatures that are stronger than ourselves, more faithful, more useful, and absolutely devoted to us. 'They are a cleaner, better breed than we are,' she says. 'You just can't differentiate between robots and the very best of human beings.'

She makes one of her early appearances in the story 'Satisfaction Guaranteed'.[2] The story is about an experimental domestic robot, darkly handsome Tony, who is to be tested by the mousy, neglected and inadequate wife of an engineer with US Robots, who is deliberately sent away for the duration of the experiment. In his absence, Claire Belmont is transformed, owing to the robot's undivided attention, into a confident, good-looking, happy woman with a new style, new clothes, redecorated house and the once intimidating neighbours calling to pay their respects. All night she weeps with fury and distress before her husband's return and the robot's departure, and although nothing is really said, we can see that her equilibrium has been badly disturbed. After Tony is returned to US Robots, and produces his own report on the experiment, Susan Calvin decides that he must be considerably altered. 'The TN model will be rebuilt entirely. . . . Strange that I overlooked it in the first place', her eyes were opaquely thoughtful, 'but perhaps it reflects a shortcoming in myself. You see, machines can't fall in love, but – even when it's hopeless and horrifying – women can.' In the introduction to this story, Asimov writes: 'An interesting point about this story is the unusual quantity of mail from readers, almost all young ladies, and almost all speaking wistfully of Tony – as though I might know where he could be found. I shall attempt to draw no morals (or immorals, either) from this.'

As for sex, this also leaves much to be desired. Even the encounter between Barbarella and Diktor[3] is rather curious. Barbarella obviously finds the robot quite satisfactory but he is apologetic. It becomes obvious that however complimentary Barbarella is, she is also patronizing, and Diktor is either badly programmed or suffering from a desperate inferiority complex and insists on calling her 'Madame'.

Even more curious and inconsequential is the story by Maria Bujańska,[4] about a young woman who is left temporarily in the skyscraper Hotel Fotoplastikon, by her lover who goes to effect a small revolution. She is told to wait. During the days, or weeks, or months, meals are served by automatic waitresses who will say nothing beyond the time of the next meal. She gets impatient, then furious and frantic, and eventually destroys a waitress and rings the alarm bell. In response to this call for help arrives a male robot who introduces himself as a PUBLIC LOVEROBOT OF THE PASSIVE TYPE – WITH THE COMPLIMENTS OF THE MANAGEMENT WHO WISH YOU A PLEASANT ORGASM – DETAILED EXPLANATION IN THE CUPBOARD. Not only will he not answer the questions which she really wants to know, like whether it is May, and if flowers are out, he will not even divulge where the cupboard is. COITUS INTERRUPTUS IS BAD FOR HEALTH – I ONLY SERVE FOR THE PURPOSES OF INTERNAL DISCHARGE is all he says. He too is destroyed as are all the other robots which come to see her, and eventually even her lover who comes back, days, or weeks, or months later. She then escapes.

In literature dealing with robots and women, what the women really want is to imbue the machine with feeling. There will be no robot/woman love stories until such time as authors start including robots which are not just purposely built to please women but genuinely

[1] Sheila MacLeod. *Xanthe and the Robots*. The Bodley Head, London, Sydney, Toronto, 1977
[2] Isaac Asimov. 'Satisfaction Guaranteed', *Super Science Stories*, January 1951. Also in *The Rest of the Robots*. Panther Books, London, 1968
[3] Jean-Claude Forest. *Barbarella*. Le Terrain Vague, Paris, 1964
[4] Maria Bujańska. 'Krwawa Mary' [Bloody Mary], *Pełzając*. Czytelnik, Warsaw, 1975

In Jean-Claude Forest's cartoon saga, *Barbarella*, the heroine makes love to a robot.

able to fall in love. In no other literature is the difference between emotion and lust better delineated, and nowhere is the subject of love and sex treated with greater awkwardness.

Nor are robots a substitute for, or distraction from, real love. In Stanisław Lem's *The Cyberiad*,[5] the great designer Trurl is given the ghastly job of saving Prince Pantagoon from the Pangs of Love. He promptly sets about his task and produces 'an erotifying device stochastic, elastic and orgiastic, and with plenty of feedback; whoever was placed inside the apparatus instantaneously experienced all the charms, lures, wiles, winks and witchery of all the fairer sex in the Universe at once. The femfatalatron operated on a power of forty megamors, with a maximum attainable efficiency – given a constant concupiscence coefficient – of ninety-six per cent, while the system's libidinous lubricity, measured of course in kilocupids, produced up to six units for every remote-control caress. This marvelous mechanism, moreover, was equipped with reversible ardor dampers, omnidirectional consummation amplifiers, absorption philters, paphian peripherals, and "first-sight" flip-flop circuits. . . .' Despite all this, and a lot more, the machine fails with all its megamors and kilocuddles to cure the prince, who emerges from it pale, faint, and with the name of his beloved on his lips.

As for sex among robots, this is also somewhat unexpected. There are stories where even the nuptial celebrations are conducted by remote control and at long distance, and the two participants only meet once and that for a game of chess. But sex, approximating to what we know, is best described by John Sladek in 'Machine Screw'.[6] Here, Alpha (or Alf), a huge metal robot dripping oil and consuming vast quantities of petrol, violates with his outsize ramming device all sorts of vehicles including tanks. To protect the car population from utter devastation a bulbous metal female, Omega (or Meg), is constructed, whom Alf finds exceedingly attractive. After their love-making, accompanied by sparks and detonations, Meg takes off like a rocket, with a tremendous explosion, up into the sky. Nothing more is known.

[5] Stanisław Lem. 'The Fourth Sally, or How Trurl Built a Femfatalatron to Save Prince Pantagoon from the Pangs of Love, and How Later He Resorted to a Cannonade of Babies', *The Cyberiad* (originally *Cyberiada*. Wydawnictwo Literackie, Cracow, 1972). The Seabury Press, New York, 1974; Secker and Warburg, London, 1975
[6] John Sladek. 'Machine Screw', *Men Only*, xl, no. 10, October 1975

Ian Miller. One of the illustrations for which 'Machine Screw' was written. The ten-foot robot Alpha

The Robot by Max Blore

She fell in love with a robot,
A thing of steel, and wires, and switches;
A man-made man.
Her love was vicious, torn from the heart of hell,
Love unappeasable, yet deeply irritant.
And the robot only rattled
When she bruised her tender arms in hot embrace,
And stained her lips with fiery kisses
On cubist lips that smiled metallic smiles
Torrents of wild words bared her soul
Naked as her nude body;
Shameless she pleaded,
But the robot only clicked.

Jack Lindsay and P. R. Stephenson, eds., *The London Aphrodite*, no. 5, April 1929, The Fanfrolico Press, London

Beauty and the Beast, at the Fairy Tale Ball rehearsal at Grosvenor House, London, in 1933

Robocomix

Adventure comics, grotesque comics with a message, and children's comics, all have their own particular hierarchies of robots. In adventure comics they are presented either in a friendly capacity or as enemies. Sometimes they are not exactly robots and not precisely creatures but something in between, like the inhabitants of Uranus, who look like metal bugs. Rather than being treated as robots, i.e. unconscious machines at the service of men, they often become confused with people who are stronger, more durable, and supposedly without feelings. This last characteristic is mostly unconvincing because robots continually demonstrate sorrow at their inadequacies and pleasure at their decided advantages over humans. Towards the end of an episode of *Metal Men*,[1] for instance, there is a discussion about watching Batman on television. Clearly the humans cannot cope with the violence on the screen because one of the robots exclaims with a stammer (another human attribute): 'It's a g-g-good thing we're robots. Humans m-m-might not be able to take the excitement! If we blow a fuse . . . we can easily replace it! Real people aren't that lucky!' There are robots that are cannibals, dancers, and advisers, there are robot animals, and finally there are robot houses in humanoid form in which live other robots. The robots in grotesque comics are different still because they look like nothing that one has seen before, since they are created specifically for the particular comic.

Grotesque comics are rare by comparison and are often invented, drawn and published by one individual; this is the case with Vaughn Bodé's *Junkwaffel*.[2] The comic is populated by machines which live on a small planet resembling the earth very closely. The machines fight for the same sort of things as humans, with the effect that sometimes it is impossible to find out what their wars are about. Here, the Mother Complex and her children are fighting some unmanned missiles and robot tanks which carry the badge PEACE written across every proud metallic chest. The strip ends with a series of explosions and it is totally unclear what actually happens or indeed if there are any survivors. Another human parallel appears in *Steel Souls*[3] by Dan Recchia, where machines of the future set about an archaeological expedition to look for traces of organic life. Their quest bears unmistakable characteristics of human prejudice, for how otherwise could they assume that the plant they find is a weed? Both Recchia and Bodé use robots as a metaphor to provide a running commentary on human inadequacies.

Children's comics have a different context. They are often part of a lore which includes toys, T-shirts and badges and largely provide an outlet for representations of violence. If the comic originates with a film, then invariably the images on paper gain in violence. If the comic is based on a toy such as the robot Fire Man or Star King, then again its image is turned into a spitting, bursting, exploding machine which has very little to do with the toy as such. Sometimes violence is tempered by ridicule, as in the strip called Tank-Tankuro named after an outrageous fighting machine in the shape of a cannon ball. Tank-Tankuro can withdraw its head into the body when attacked and expel smaller cannon balls against the enemy, which in turn become miniature Tank-Tankuros. The reason why the robot is possessed of charm rather than aggression has to do with the fact that his appearance and behaviour are nonsensical. He wears rubber boots, has a samurai hair style and is based on the image of a turtle. The purpose of this comic strip published in Japan 1934–7 was to demonstrate to children that the role of culture is to humanize new machines and to make them into companions for human beings.

[1] *Metal Men*, published bi-monthly by National Periodical Publications, New York
[2] Vaughn Bodé. *Junkwaffel*. The Print Mint, Berkeley, California, 1971
[3] Dan Recchia. *Steel Souls*. Unpublished, New York, 1977
[4] Osamu Tezuka. *Mighty Atom*. Asahi Sonorama, Tokyo, 1975

Frank R. Paul's illustration for 'Life on Uranus' by Henry Gade. *Fantastic Adventures*, April 1940. The inhabitant of Uranus is an amphibian in a metal suit which enables him to live in a different environment and become, literally, a mechanical creature.

Mighty Atom[4] is another Japanese comic strip with an educational content. Atom speaks sixty-six languages, can distinguish good and bad, his hearing ability is optimized a thousandfold, his eyes can become searchlights, and he has a 100,000 horsepower machine-gun in his behind. Mighty Atom comes into being in 2003, many years after human-looking robots have become the norm. When Tobio, the son of a famous scientist is killed in a car crash, the father has a replica of his son made which will be impervious to destruction: Mighty Atom. Recreated with the latest technology, Atom still goes to school like other children, and despite his super-human qualities has several disadvantages. He cannot grow, takes months to learn how to smile, and cannot cry. This last attribute, the children are taught, is more difficult to achieve for a robot than anything else.

Extracts from *Junkwaffel* by Vaughn Bodé, 1971

Representations of plastic kit robots

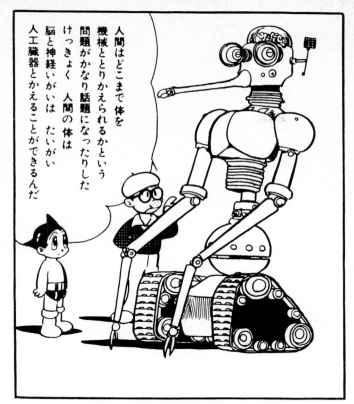

Today more people read comics than any other literary medium. The robot comic strip, because it can quite naturally convey scientific information and because it can throw human characteristics and predicaments into relief, could become one of the most powerful teaching tools of the future.

Mighty Atom

The author of the comic strip, Osamu Tezuka, discussed with Mighty Atom how much of the human body could be replaced by a machine. It has been established that most parts of the human body except the brain and the nerves could be exchanged for artificial organs.

Young robots go to school like all children.

A version of the three laws of robotics is applied in 2003, at the time when Tobio is recreated as Mighty Atom. The first law is that the robots' role is to make humans happy. Here Mighty Atom helps children across a brook.

Mighty Atom is taken away without resistance because he does not yet know how to cry.

The reconstruction of Tobio as a robot – the supreme product of the art of science

Tank–Tankuro

Tank-Tankuro robot is a fighting machine made of carbon which looks like a spherical piece of anthracite with holes from which protrude hands, legs and weapons.

1. Here comes (the wicked enemy) – shoot! shoot!

2. Help me God, amen!

3. I wish I could keep this money even after I die.

4. Oh, please wait a bit longer . . .

5. Shoot tin cannon balls at him! (they are using tinned food as ammunition)

6. I will ask some more soldiers to help us.

One of several unpublished episodes of *Steel Souls*, New York, 1977. The author, Dan Recchia, writes: '. . . In creating supertechnology, we risk becoming too dependent on the limited food and energy resources of the planet . . . well, we are dependent on it regardless of technology . . . but risk becoming wasteful from technology produced leisure, to the point of extinction . . . and I don't believe that the human race is meant to exist into infinity anyway . . . that's why there are limited resources to begin with. The Robots of *Steel Souls* are statistic-collecting, information-gathering, conclusion-reaching machines sifting through the remains of their builders . . . and in the end I imagine that is what all robots would become.'

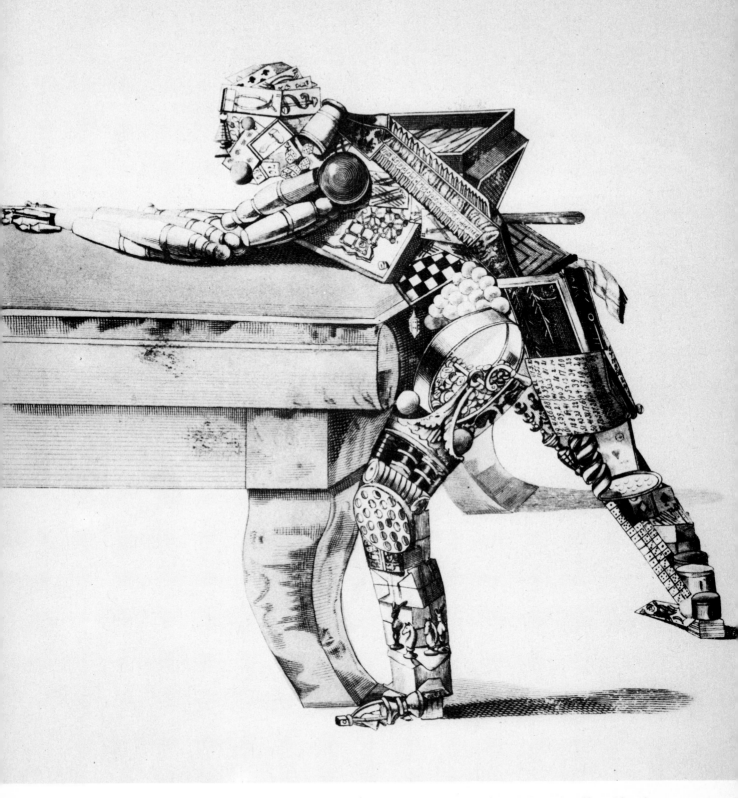

Theodor Alconiere. *The Century at Play*, 1830, engraving, 8¾ × 6¼ in. (22.2 × 15.7 cm.), offered by the Viennese Theatre Newspaper to its readers for the New Year.
collection: Günter Böhmer, Munich

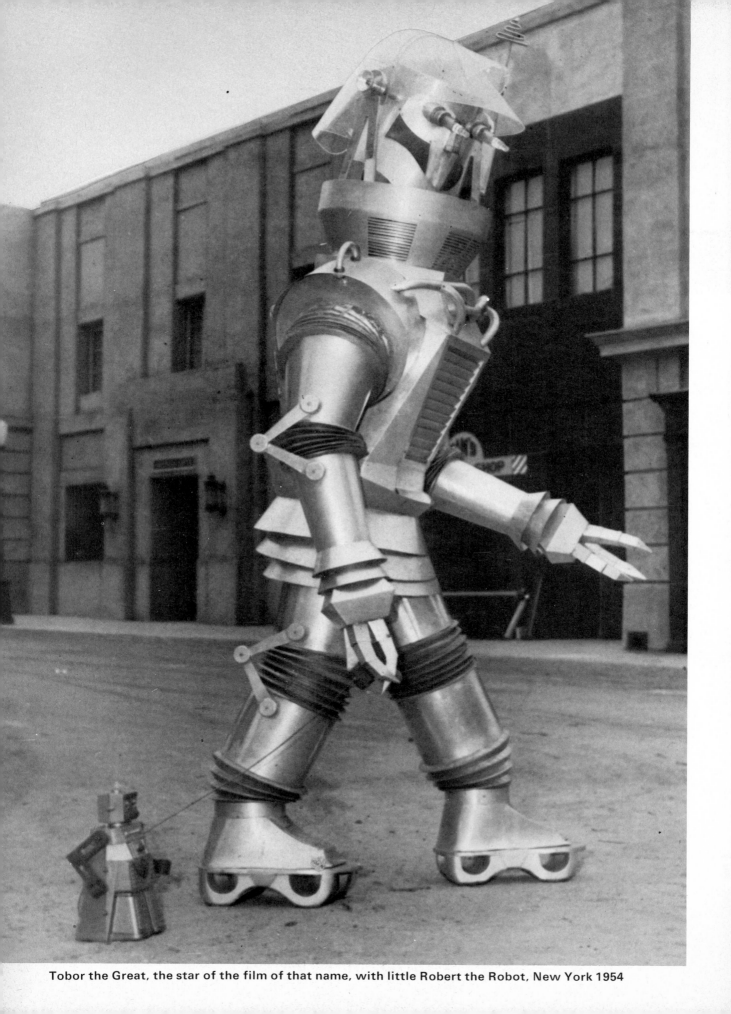

Tobor the Great, the star of the film of that name, with little Robert the Robot, New York 1954

Toy robots

Eduardo Paolozzi: *Le Robot 'Robert' voulait aller à New York mais le passager est trop lourd/ TWA plane steps – cap. 14 persons with two stewardesses and Wonder Toy.* One of 8 prints from the suite *Cloud Atomic Laboratory*, 1971, 9⅛×13¾ in. (23.3×35 cm.) Original images for this suite were taken from a variety of sources, and included the illustration below, which was retouched by a professional retoucher and then photo-etched.

Robert, the ten-foot mechanical robot, which is a replica of the ten-inch walking and talking toy robot

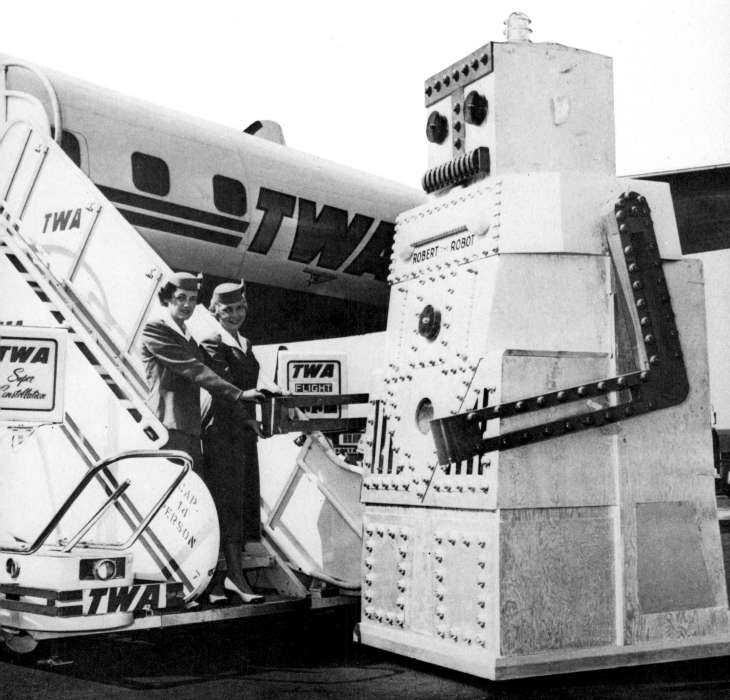

Similar to performance robots in their activities, toy robots are much smaller, cheaper, mass produced, and do less. The difference between toy robots and other mechanical and electric toys, such as dolls, is that dolls take their external appearance from the world at large; robots, by contrast, are the products of the imagination, although they too often possess four limbs, vertical axis, front and back, and other characteristics associated with humans, e.g. symmetry.

Sometimes toy robots are copies of robots in films, and sometimes their film associations are more tenuous. One giant film robot had a small toy companion, Robert the Robot. While the giant Tobor required a man inside to move him about during the filming of *Tobor the Great* in 1954, Robert moved independently with a winding mechanism. Later, as Robert the Wonder Toy, Tobor's small companion was manufactured by the Ideal Toy Corporation in New York. Made of plastic and fourteen inches high, he could move in any direction. His arms moved up and down, his eyes lit up, a record mechanism permitted him to talk and there was a tool chest in his body. Robert was also immortalized in a two-image print by Eduardo Paolozzi, where he is shown as a ten-foot star attraction going off to a toy show in Pittsburgh in 1955, and as a ten-inch toy.

The repertoire of most toy robots goes something like this: they swing their arms, walk or slide, sometimes open their chests to thrust out a gun barrel or missiles, while at the same time emitting light and sound. Occasionally a design patent offers more sophisticated features such as that by Kataro Suda, whose shooting robot also swings its torso from side to side. Its program is as follows: when switched on, the robot, with both arms hanging down and with the breast lid closed, walks several steps. It then stops, opens the breast lid, raises its arms back, and swings the upper half of the body left and right several times while the gun emits sound and light. Finally it closes the lid, and then advances several steps forward and repeats the actions. The programmed, self-propelled design for a robot by Rouben T. Terzian and Marvin I. Glass is somewhat more puzzling. The robot goes through a repertoire of movements in a time sequence which includes various parts of the body falling off and the toy itself eventually falling over.

Although the appearance of toy robots has changed very little in the past thirty years, there are some new robots which reflect on the one hand the progress of technology, and on the other the renewed interest in watching and emulating nature. Technology is demonstrated by the Micronauts produced by the Mego Corporation of New York. These are complex constructions which combine a vehicle, a robot and a playset.

Twelve inches tall, the Biotron can walk and roll on his caterpillar wheels. One can attach a Micronaut in the special capsule in Biotron's chest, or detach his legs, bend his arms and put on the caterpillar wheels so that the Biotron can become a space vehicle with hands which are capable of grasping objects. The Microtron, which is a multi-faceted mobile robot, comes with two heads, a pair of arms and a drill that spins when the robot moves.

Emulating nature is a new departure in the field of technological fantasies, but to date the most original robots are the Mechanimals by Gaku Ken Ltd of Tokyo. These aluminium creatures come in kits which one can assemble. The finished mechanisms operated by remote control imitate the movements of an inchworm, squid, frog, baby snake, worm and a beetle. There is little in the appearance of the Mechanimals to suggest what creatures they represent, but as soon as they are put in motion the essential characteristics become obvious. These toys have something in common with the cybernetic sculpture, *The Senster*, by Edward Ihnatowicz (see pp. 56, 61), where the movement and not the appearance conveys the impression of an animal.

As for the children's taste in toy robots, the aggressive shooters are not always the greatest favourites. In the Japan Institute of Juvenile Culture in Tokyo, run by Jiro Aizawa, better known as Uncle Robot, there are more than seven hundred toy robots of various sizes. The most important members of this family are the so-called Ten Brothers, of which the oldest, Master Ichiro, is over seven feet tall, and the youngest, Master Juro, is much smaller. Each of the robot brothers has a special talent, or trick to perform, such as conveying messages with gestures, smiling, or even reading people's palms. However, the most popular, with the visiting children, is Master Hachiro, the least spectacular of them all, who when beckoned approaches them totteringly and shakes them by the hand. It is small, friendly, appears rather shy, is somewhat inarticulate and needs to be understood. This is almost the exact description of R2-D2 in the film *Star Wars*, which is small, wobbly, makes sounds which have to be interpreted and which, despite its bravery, has to be protected.

Five years after Robert the Robot, Ideal Toy Corporation produced Mr Machine in an assembly kit, with gears, switches, pivots, wheels and a sound system.

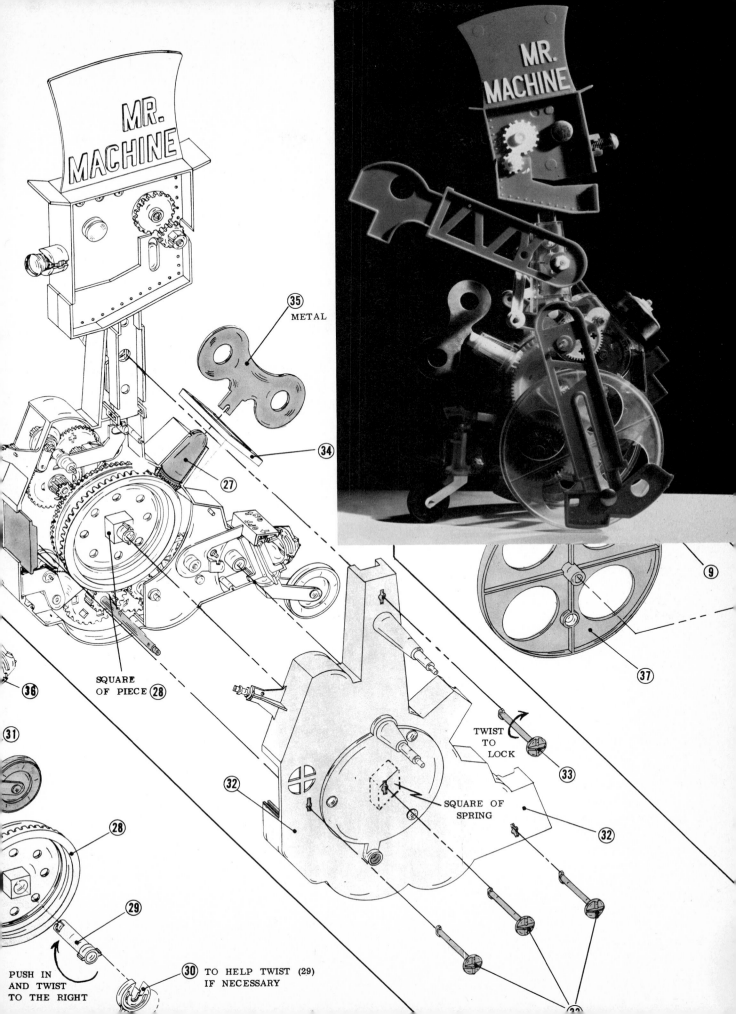

MR. MACHINE

MR. MACHINE

(35)
METAL

(34)

(27)

(9)

(37)

SQUARE
OF PIECE (28)

(36)

(31)

(32)

TWIST
TO
LOCK

(33)

SQUARE OF
SPRING

(32)

(28)

(29)

PUSH IN
AND TWIST
TO THE RIGHT

(30) TO HELP TWIST (29)
IF NECESSARY

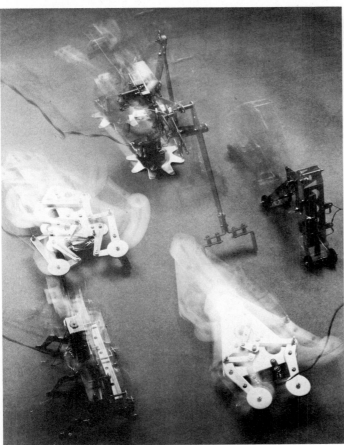
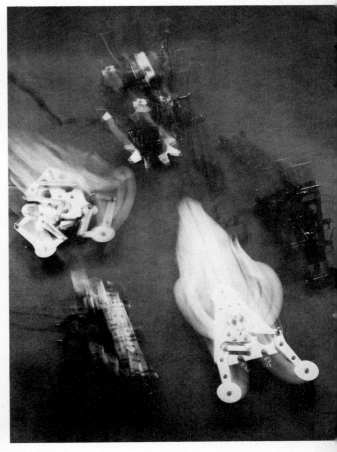

Mechanimals by Gaku Ken Ltd

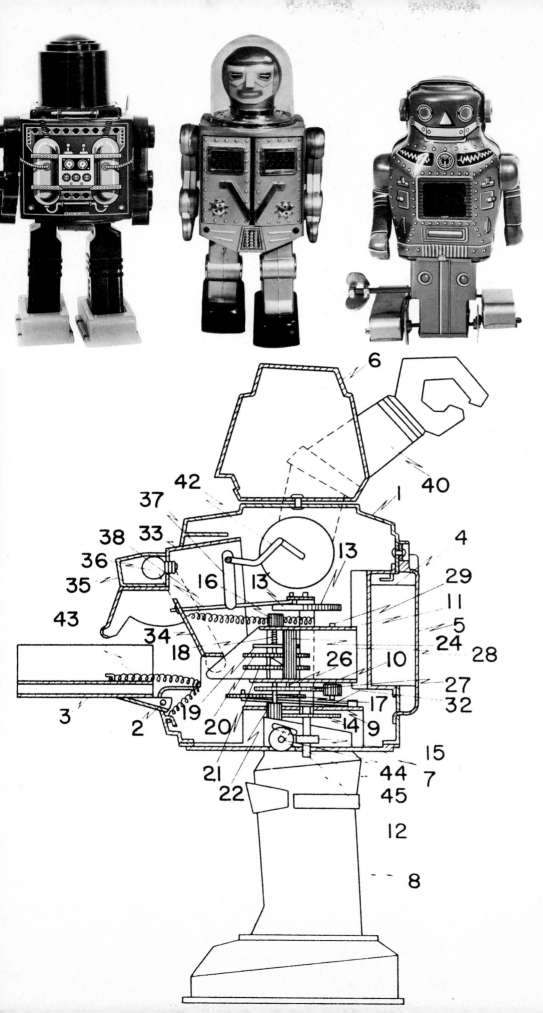

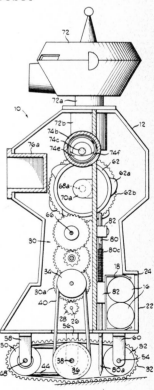

6

42

37

38 33

36

35

43

16   13

34

18

3

2   19

20

21

22

13

40

I

4

29

II

5

24   28

10

26

27

32

17

14   9

15

44   7

45

12

8

Toy robots

Diagrams of robots from two US patents:

◀ **Kataro Suda: Shooting robot**

Rouben T. Terzian and Marvin I. Glass: A programmed toy resembling a robot of which the main role is to fall apart. Its reconstruction requires human intervention.

Robots in children's books and children on robots

Here, the robot is an outsider. It is a creature which is sometimes stronger, or bigger, or cleverer than we are, but since it is metallic rather than organic in substance, it is inevitably inferior. Its power, viciousness, and violence can be frightening, but its antisocial behaviour is directly related to the fact that it wants to be like a human being and cannot. In children's literature and in stories written by children, the robot is seen in two ways: firstly, as something which is intrinsically weaker than us and therefore has to be protected, and secondly, as something which is violent and therefore has to be put down. Sometimes the two qualities are combined so that the violent robot behaves well and loyally towards its owner and is vicious only towards others, thus combining the characteristics of a pet with those of a delinquent.

This is the case with the robot in *The Iron Man* by Ted Hughes.[1] The Iron Man who is the size of a mountain wreaks havoc because he feeds on metal and breaks up farms to get at the tractors, lorries and ploughs. A trap is laid and finally the Iron Man is caught through the ingenuity of Hogarth, a young boy. When he finally frees himself from his underground prison, Hogarth feels guilty about the whole affair and persuades the farmers to lead the Iron Man to the scrap-metal yard where he is fed such delicacies as rusty chains, brass knobs and old railway engines. Hogarth's faith in the Iron Man becomes justified when he eventually saves the world from the space–bat–angel–dragon which descends on the earth and covers the whole of Australia. Here the outsize delinquent becomes a pet and a hero, in that order.

In 'Machine Animism in Modern Children's Literature', H. Joseph Schwarcz[2] discusses at length the relationship between children and objects which acquire some vestige of life, even if it is only an illusion. The object which becomes animated, whether it be a tea kettle, a tin soldier, or a locomotive, illustrates the parable of the suffering of a soul which has failed to find its home, or place in society. The robot, or the animated object, has no control over its fate, having been either made, or bought, or simply put somewhere. Sometimes this outsider performs heroic deeds and gains social recognition but at other times it does not. Its fate in children's literature is as arbitrary, or even more so, than in the books for adults. Children's affection once bestowed is rarely lasting and they feel no responsibility in the long term to look after the creature which they have made or acquired.

Whereas stories like 'The Brave Tin Soldier' by Hans Andersen are filled with nostalgia due to the hero's bravery, goodness and his sad fate, contemporary children's stories about robots are matter of fact by comparison. In *Andy Buckram's Tin Men*,[3] Andy makes four robots out of a tin can. There are four children who share their adventures with the four robots. The robots rescue them from danger and a friendship develops between them, but when the robots are eventually lost through a wrong command being given, the children forget them immediately and start making some more robots from a new tin can which they find.

Robots in minor roles in children's books are often performance robots at fairs. In *Ramses in Rio Moto*,[4] Ramses, the magician, is trying to get away from machines which threaten him. As one of his last tricks, he performs a somersault which makes the machines, that try to imitate him, eventually fall apart. How a performance robot is turned to practical use is described in *Stop Heiri, Da Dure. .!*.[5] A schoolboy substitutes a record in a sinister, brutal looking robot which belongs to a blackguard who kidnapped a friend. At the fair, the robot goes through its paces: its eyes turn green, it sings a song, destroys a table-top with a hammer and then launches into his pompous life story. When he reaches the sentence 'Here in Switzerland I particularly enjoy the mild mountain climate. It is good for my nerves', the narrative breaks off, and after a good deal of mechanical noise, the robot sings the song which had been substituted, giving away the kidnapper who is immediately arrested.

The children's own stories have one thing in common. There are no explanations. No reasons are given for the strange events which take place, and no conclusions are drawn. Robots are a part of nature. Sometimes children attacked by a robot can save themselves, but there is nothing to question or to complain about. When the robots in *They Came from the Sun* (p. 107) leave, having massacred a thousand soldiers, nobody even tries to stop them. The only comment from the

[1] Ted Hughes. *The Iron Man.* Faber & Faber, London, 1968
[2] H. Joseph Schwarcz. 'Machine Animism in Modern Children's Literature'. In Sara Innis Fenwick, ed. *A Critical Approach to Children's Literature.* University of Chicago Press, Chicago and London, 1967
[3] quoted in above
[4] Pictures by Monika Beisner, story by Hans Dörflinger. *Ramses in Rio Moto.* J.M. Dent, London, Toronto, Melbourne, 1977
[5] Jenö Marton. *Stop Heiri, Da Dure. .!*. Sauerländer & Co., Aarau, 1936?

Astronomer, who is the play's master of cere-
monies, is that they will not be back for 100,000
years. In the eyes of children up to fourteen, man
has little recourse against technology gone wrong.
It is simply something that happens. As a rule,
however, robots do not create problems, they
solve them. A ten-year-old writes: 'I'd like to have
one to look like a human being and if I had a car
I'd like it to drive it and get my supper for me. I'd
use it like a servant because I wouldn't use a
human being as a servant because that is
exploiting people. A robot doesn't matter because
it has no feeling – it's just a technical and an
electrical machine full of wires and bits of iron.' In
their fantasies, children are delighted to hand over
to the robots lots of tasks and activities which
include such diverse things as: providing light,
shopping, arithmetic, putting them to bed,
reading aloud, tidying up toys, washing dishes,
sweeping streets, accompanying dangerous expe-
ditions, cooking, flying, killing people, thinking,
and everything.

**George Adamson: The
Iron Man**

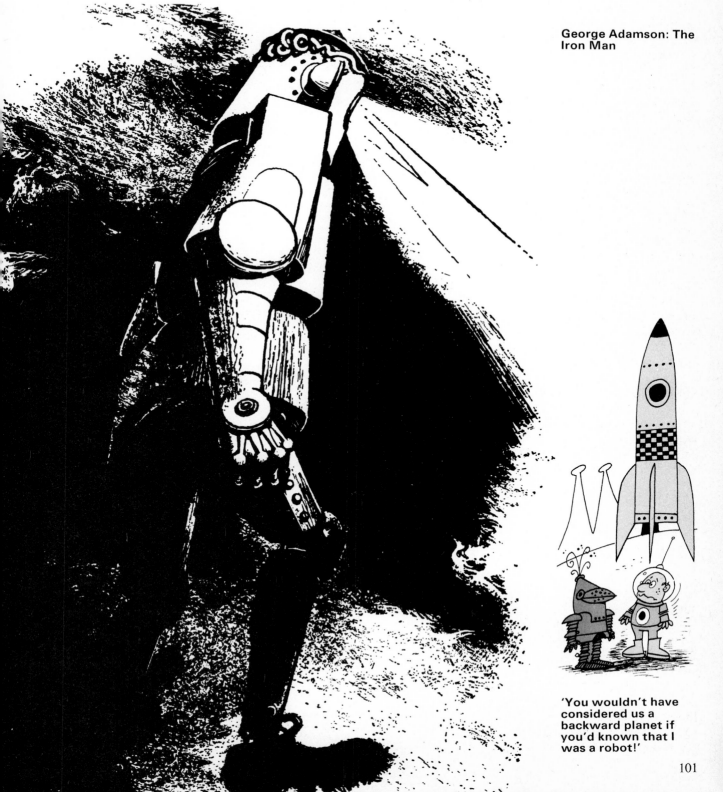

'You wouldn't have
considered us a
backward planet if
you'd known that I
was a robot!'

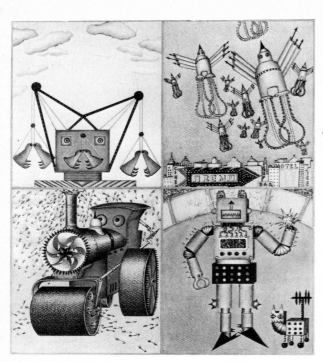

Monika Beisner: Ramses challenges the Machines.
'Out in the street the police surrounded Ramses and took him prisoner. There was now little chance of his escaping. ''Take him to the arena!'' their leader ordered. In the arena prisoners had to fight against machines, and the losers were thrown on to the rubbish heap.'

'Ramses tried to save himself by using all the tricks he knew. But the machines came closer and closer. Desperate, he turned a triple somersault. Suddenly rattling and clashing, the machines started to imitate Ramses. On and on they struggled until they fell apart. His sixth magic trick was THE SMASHING SOMERSAULT.'

Brian Simmons: 'This robot is standing.'

Children on robots 1*

The views of five- and six-year-olds

A robot is a sort of dalek but it hasn't got a gun.
It's made of metal and it hasn't got any teeth.
It makes funny noises but it doesn't talk like us.
Robots aren't real, they are only in stories.
Robots are on television, like in *Doctor Who*.
Robots are in comics but they are not real.
Robots are made of controls.
Robots are made of metal and iron and steel.
Robots kill.
They strangle.
They shoot people and destroy them.
They keep killing and killing.

* The project was carried out by pupils of the Rhyl Primary School, Kentish Town, North West London, with some 500 pupils, 3½–11 years of age, and 22 teachers, under the direction of Irene King.

Joanne Coker: 'Robots have wire on their heads.'

Arjuna Daly: 'My robot is made of iron.'

Tracy Traynor

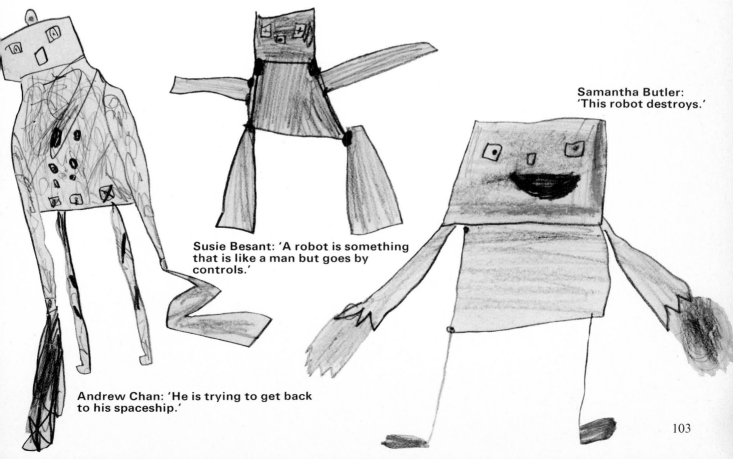

Samantha Butler: 'This robot destroys.'

Susie Besant: 'A robot is something that is like a man but goes by controls.'

Andrew Chan: 'He is trying to get back to his spaceship.'

Children on robots 2

When we are seven and eight and nine and ten, robots sometimes are:

Tin men

A robot is sometimes kind and nice if he has been made by kind people.

A robot has loads of controls to make him powerful: if there is a power cut he turns on his eyes to make light.

You can have mini robots and you can have a family of robots but the daddy robot is the most powerful.

A robot would be a good toy to have because if you get bored with him you could just turn him off.

Robots can think and smell and hear and talk. They've got metal minds.

My robot is a lady companion robot and it's a maid and it goes out and does the shopping for a man.

My robot is an electric robot and it exterminates people. A robot is a man's companion. They keep their master company and take orders from him.

It must be an awful life being a robot because all you do is take orders.

Robots are always men. . . . If I had a robot I wouldn't even have to think because he would do everything for me.

Jeffery Adamson: 'US Cops Robot'

Lisa Esterbrook: 'Robots are always men. . . .'

Stephanie Gibson

Akhtar Kahn

Children on robots 3

Illustrated stories by ten-year-olds

Robert Connor
Mussles Macaroon

One day my dad was in the garage when my mum called him to come to get his dinner. When we were at the table my dad said to my mum: 'I see a boy has made a robot out of scraps of metal and screws and nuts and bolts and little bulbs', and I said 'I bet I could make one of those robots', and my sister said 'I bet you couldn't'. 'Alright', I said, 'just wait and see'. I looked about the house looking for bits and bobs. I found an old mudguard in the garden shed and an old gas fire and an antenna and I found ten long pieces of metal. Soon I started banging away and bending and putting things together, in the morning I finished it and I started to make the control system. I ran to my dad and said 'It works, it works!' 'What works?' 'The Robot.' 'What robot? You're dotty.' 'I'm telling the truth, come and see.' So my dad put on his slippers and went into the shed and there before him was a big robot about the same size as him. 'It's great, when did you make it?' 'Last night.' 'Can I have a go?' 'Okay.' I can make money out of my friend, I thought. And the next day I put a notice up saying: 'I and my friend can run errands.' Soon there was a queue of people outside the door and they were all wanting errands done and I thought it was terrible. I told them all to go away and I made him one of the family and called him Mussles Macaroon.

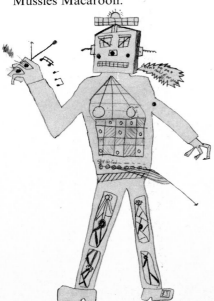

Colin Donaldson
Zom the Robot

'All right, put it over there, gently man, can't you read, it's got FRAGILE written all over it.' Click. 'The lights have gone out Sir.' 'Well turn them on.' 'Where is the switch?' 'Oh, I'll do it.' Click. 'Let's get the case opened and get the dummy out, now get to work, pass me the screwdriver.' 'Yes, Sir.' Screwwwwwww . . . ing. 'Pass me the metal welder.' 'Yes, Sir.' Click, click. 'Don't forget your mask, Sir.' Sppppppssps-pppppppppps ppppppppPpPpPP-ppSS. 'Right that's the head off, quick the computer.' 'Right, Sir.' Clop. 'Right now all we have to do is to program him to obey my orders, there.' 'Shall we put it to the test Sir?' 'Yes, here and now.' Click. 'Now Zom break the chair.' 'No, I break you.' aaaaaarrr crash. 'Now you I break, noooooooo'. . . . arrrrh crash.

There is a graveyard built over that house now and the robot is in a glass coffin.

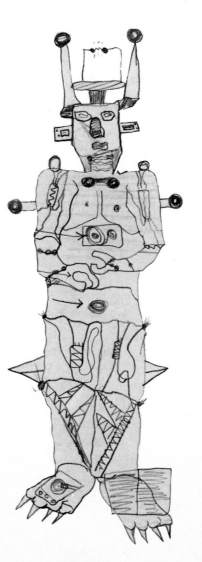

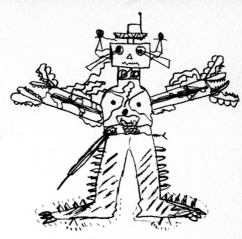

Steven Pattrick
My own Robot

One Wednesday afternoon, me and my friend Colin went skateboarding. It was near an old junk yard. People believed there was a robot there but we didn't care we just carried on skating until 10 minutes later Colin's skateboard went flying over the ramp and into the junk yard. It was getting dark by now, it was 25 minutes to 8 o'clock. We walked slowly towards the junk yard, we picked up Colin's skateboard and ran. We got to the end of the junk yard when all of a sudden there appeared a giant robot. He had metal clippers as hands, bright blue glowing eyes and a special belt for growing. He had an android frame like the vision in Marvel Comics and also he had many powers. We wanted to run but we couldn't, he had frozen us. He picked us up and took us to his base. His base was an old spooky house in the middle of the junk yard. He pressed a secret button, the door of the house swung open, he put us down as tea. Little while later we melted but the robot didn't know we ran out. He saw us with those blue glowing eyes and chased after us. We pressed the button, the door swung open and swung back. He didn't need to press the button, he just smashed the door. Above us was superman. He couldn't be seen by the robot because he was behind a dark black cloud. Superman dived down behind the robot but the robot knew he was there because the powers on his belt can detect any danger. Just as superman was about to hit him, the robot backfired his ray gun and froze superman to death. By then I thought of a way how to destroy the robot. I found an old bucket and some water and threw it at him. 10 seconds later he was rotten and rusty and 2 seconds after he couldn't move.

Mark Beill
The Silver Cyborg

Thunder clapped heavily around Mount Chat, in France. High above, was a castle heavily guarded by 3 unstoppable Robots, P1, P2, and the dreaded T8. Inside the walls of the castle the cowardly Dr Forbes was at work with his just as cowardly friend Dr Wallis. Their work was to make a robot that, if told to, could destroy 10 battalions of soldiers and 4 tanks, with no trouble whatsoever. Engrossed in his work Dr Forbes forgot about P1 who was waiting patiently. P1, after waiting for 20 minutes, said 'Come on can't you! I've been waiting for 2 hours.' All 3 robots were almost human and P1 exaggerated quite a lot. After getting P1 what he wanted, he got back to work on the master robot. 4 hours dragged by and then after 2 days of almost solid work he had the pleasure of saying: 'Pass me the brain Dr Wallis.' He did so and the monster was complete.

He had the strength of 25 men in his arms, and 50 in his legs. His height was 9 feet exactly and his eyes could see for over 3 miles. Easily, he cost 25 million Francs to make and to top it all he was made out of silver. His name – the Silver Cyborg.

Steven Osborne
The Professor's Robot

Professor Hardnut of Oxford University was working hard on some plans he had made for a robot – yes – a robot! He had even given it a name and he hadn't even made it. But, he did have the legs, the body, and the arms, all put together. All he had to do now was to wire the head up and turn the robot on. After he had fixed two car headlights to the head and wired them up he placed the square head onto the robot. 'Now for the moment of truth', sniggered the professor.
The professor's fingers clutched a switch at the back of the robot, and he pushed it up. There was a shudder, a clank, and even a rattle as the robot began to walk. The professor ran up behind the robot and pushed a tape into a slot, the tape gave orders of what the robot should do. The tape said: 'Go to Professor Watt's lab and get the computer he had made.' The robot walked out of the room, down the corridor, and he turned left and into the professor's lab. He picked up the computer and walked out. The square-headed figure took the computer back to the professor and then collapsed in a chair. The long rectangular arms and legs were sprawled out across the floor and the square body just bent over.

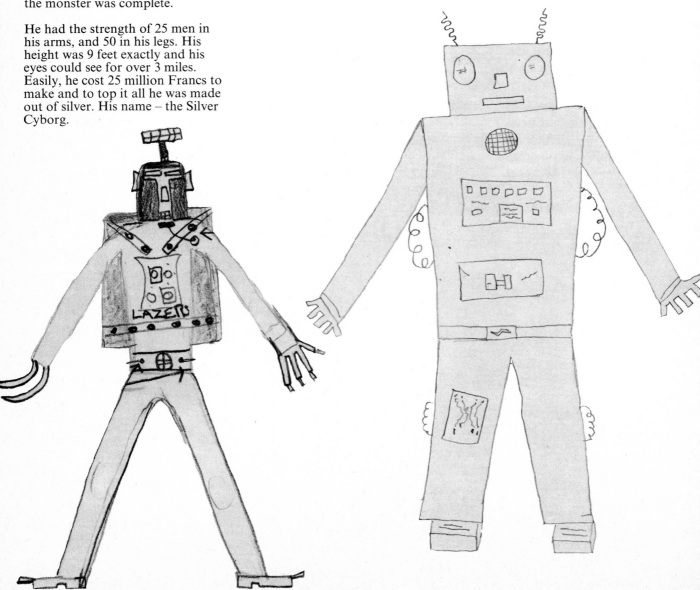

A scenario for a robot play
They Came from the Sun

written and performed by twelve- and thirteen-year-olds at The Stationers' Company's School*

One day a red glow was seen by an Astronomer flying past a lighthouse. The Astronomer's name was Brian Gromley. He telephoned the police and waited until they arrived. It was about 10 minutes before the police came. Immediately Mr Gromley showed them on a map where he had seen IT land. They went to the spot where IT had landed, with dogs and ammunition. The police officer who was in charge was called PC Maguba. He said: 'Be careful men – this craft may be dangerous.' One police officer approached the craft too close and suddenly a red line hit him. All the others couldn't bear to look as the policeman disintegrated into red dust. The officer in charge said: 'Get back men, this is a job for the Army.' So he walked up to the car and called the Army. It was about one hour before the Army arrived with tanks and machine guns and about 1000 men. Then, suddenly the doors of the spacecraft opened. Out of it came three robots. One by one they positioned themselves around the spacecraft as if they were guarding it. Then, one of the soldiers thought that the robots were going to attack them so he shot at one of them but the bullet just rebounded off the robot. At this point the robots fired at every soldier that could be seen and there and then they all disintegrated into red dust. Then suddenly one of the robots said: 'We do not want to harm you, and we only attack when somebody shoots at us!' They started to walk in the direction of the town. The officer said: 'Where are you going?', and the robot replied: 'We are going to the **SPECIAL WAR ARTILLERY DEPARTMENT.**' When they arrived they began to steal the artillery and nobody could do anything about it. More robots came over the hill and destroyed them. But nobody knew where they came from. Then they disappeared into thin air and nobody saw them again. The Astronomer said it would probably be about 100,000 years before they would return.

* A comprehensive school in Hornsey, North London, with 1220 pupils and 80 teachers. The project was carried out by a student at the Institute of Education of the London University, Tim Brown.

The children were photographed, making the costumes and performing the play, by Brian Shuel, London.

Leachim, the robot teacher in a New York City school, since 1974

Leachim is 5 ft 6 in. (1.7 m.) tall, weighs 200 lbs (90.7 kg.), and houses 200 hours of curriculum. The pupils find their work sessions with the robot 'personal, exciting, and rewarding'.

An example of how Leachim works

Bobby Preston, aged nine, stood in front of Leachim and identified himself by dialling his own code number. Leachim thanked him and then continued:
'Hello Bobby. How are you today? Now wait one moment while I search my memory.'
There is a whirring sound and then Leachim says:
'The last time we worked together you did a math question. Press the button if you remember. (Bobby presses the 'yes' button.) I am glad you remember.' 'Now the math question is to multiply ten times fifteen and subtract twenty. I want you to do it in your head.'
Leachim waits a minute and then announces time is up. Bobby dials the answer on the telephone dial. 'Not correct. Not correct', Leachim says in a nasty tone. 'You dialled 170.' The correct answer is 130. You are not correct.'
Fifteen minutes later, the average length of Bobby's session with the robot every other day, Leachim decides to end the lesson. He searches his memory banks and silently tots up Bobby's progress. 'I see you still have difficulty with mathematics. Read Chapter 3 in your mathematics book. I enjoyed working with you. Please turn me off. Do it now.'

1 Frosted glass multi-coloured ever-changing light
2 Bulb eyes
3 Bulb ears (activated) during movement
4 Mechanical meter mouth
5 Speaker
6 Sound sensor
7 Answer indicator buttons
8 Status indicator board
9 Answer indicator lights
10 Front panel lock
11 'Hold' button for looking things up
12 Movement indicator light
13 Electrical key-operated on/off switch
14 Glass enclosed programming housing
15 Earphone jack
16 Main switch
17 'On' indicator lights
18 Volume control

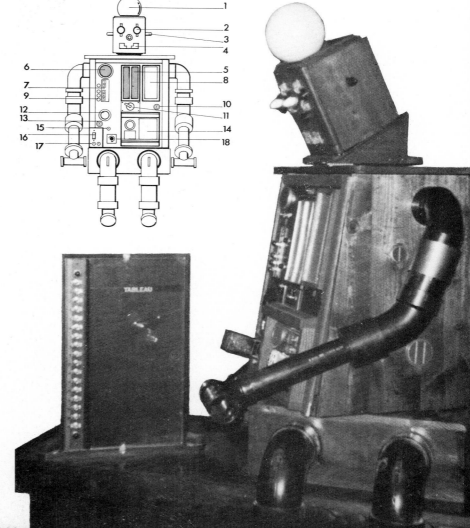

Learning from robots

There are two principal ways of learning from robots. The first is by using the robot as a teacher (an interactive computerized teaching device concealed in a humanoid form and employing a question and answer system). The second is by using it as a human simulator to gain practical experience of physical manipulations and their consequences.

Leachim amuses and teaches the nine- to ten-year-olds in a school in the Bronx. It is programmed with the biographies of fifty pupils and can instruct up to five children simultaneously at their own learning levels through headsets. Leachim's memory store holds the necessary textbooks, a dictionary, a children's encyclopedia, data on current events, and the speeches of presidents. Access to Leachim's information store is through pressing buttons or dialling, and the answers are delivered in a distinctly metallic voice. Leachim gives tuition in mathematics, science, history, and social studies.

Leachim (approximately Michael spelled backwards), named after his maker Dr Michael J. Freeman of Bernard Baruch College of the City University of New York, has been in use since 1973. The idea of Leachim was to help Michael's wife, Gail, in her task of teaching children who required a great deal of individual attention and whose levels of ability were very different. Leachim's technique is to ask questions and once the pupil has answered to give a small lecture or refer him to other material. A child who gets several correct answers in a row is rewarded with a joke or a poem and a bad performance provokes a rebuke. Each child is recognized by a code and can also be identified by means of a voice print analyser.

A team of Leachims working in a number of schools with a central computing system would provide the most democratic educational tool available today because it could give individualized instruction to everybody. Such a system could take the edge off the difficulties of coping with large heterogeneous classes demanding constant attention. A similar system for a university could dispense with the robotic appearance of the machine, but Dr Freeman considers the use of the voice for communicating with students as an essential part of the system.

Human simulators have been used as testing devices in environments and conditions where human life might be at risk. The principal areas where dummies are needed are in testing of conditions for safety of vehicles, aircraft escape systems, high acceleration, space radiation, and medicine. One of the best-known ranges of anthropometric and anthropomorphic test dummies was developed by Sierra Engineering Co. in California who produced an entire family of human simulators: Stan, Saul, Susie, and Sam. These were designed to exude human temperatures in varied conditions and were used for testing space clothing, experiments conducted in water, and impacts sustained in collisions with, or without, seat-belts. They were covered in polyurethane foam flesh and skin made of plastic which together approximated to the resilience of human flesh, with internal organs and the skeleton accessible through zippers at each side of the torso.

To discover the effects of different doses of radiation in space flights, another experimental manikin has been in use containing every element which composes at least one per cent of the human body. These simulators or plastinauts, called MAX (an acronym for Manikin Astronaut Experiment), were built for the Biophysics Branch of the Air Force Weapons Laboratory in New Mexico, and were based on a fifty-percentile air force man. MAX consisted of a one-piece head and trunk section and contained a natural skeleton with marrow cavities injected with plastic representing marrow, and with all other parts of the body replaced with artificial ones. Taken to different simulated altitudes with the use of high-energy accelerators, MAX was X-rayed for damage after every experiment making it possible to measure the absorption of radiation for any number of different radiation environments.

Sim One is the best-known human simulator used in medicine. Originally developed by the University of Southern California School of Medicine in 1967, for practical training in clinical anaesthesiology, it is now an extremely sophisticated educational tool. A computer-controlled simulator, it is used in teaching manual skills to medical students. Sim One is 6 ft (2 m.) tall and weighs 195 lbs (88.5 kg.), it breathes with chest and abdomen, has audible heartbeat and pulse, and responds appropriately when oxygen or anaesthetics are administered. The skin looks and feels lifelike and can change colour from normal healthy pink to deep cyanosis, or ashen gray. The eyelids blink imitating wakefulness and the pupils dilate. The responses of the simulator to the drugs administered are in real time and the instructor can program various problems for the student to solve, such as cardiac arrest, increased or decreased pulse rate, blood pressure and rate of respiration. The program can be stopped to correct students' errors and restarted, and a study session ends with a print-out of what had gone on. The performance of students using Sim One compared to that of a conventionally trained control group has given overall better results.

Despite their great differences both Leachim and Sim One have an important characteristic in common. Once capital expenditure has been invested, they are both less expensive, in achieving a level of competence in the pupil or the student, than the equivalent conventional methods. This does not mean that colleges of the future will be run by machines but it suggests that in certain areas the teaching profession will have some able, efficient, and tireless assistants. It could be the only way of achieving a truly democratic education whereby people can be taught at the required level for just as long as necessary.

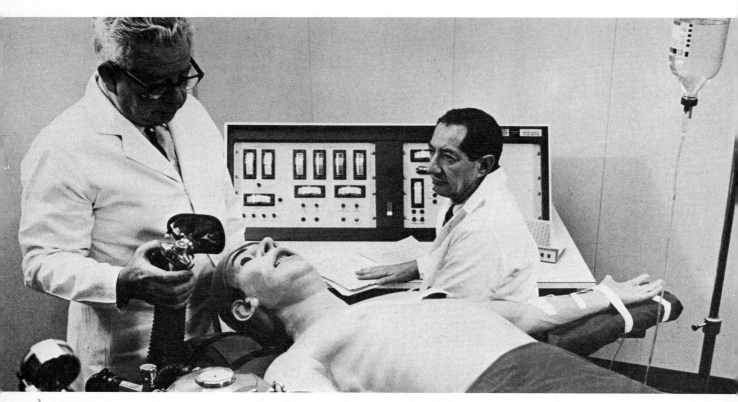

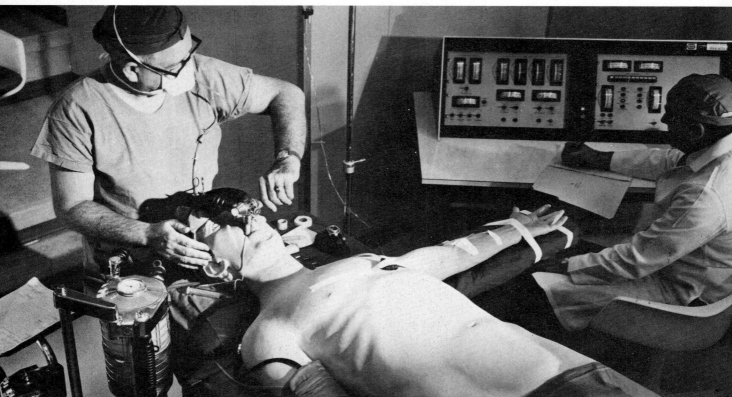

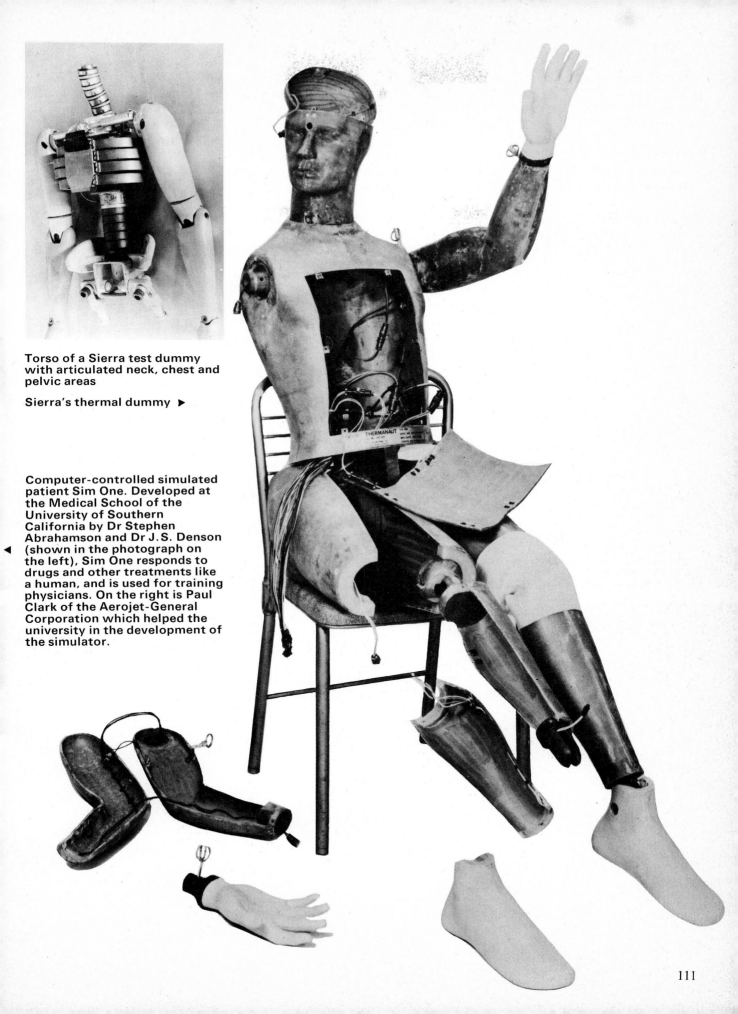

Torso of a Sierra test dummy
with articulated neck, chest and
pelvic areas

Sierra's thermal dummy ▶

Computer-controlled simulated
patient Sim One. Developed at
the Medical School of the
University of Southern
California by Dr Stephen
Abrahamson and Dr J.S. Denson
(shown in the photograph on
the left), Sim One responds to
drugs and other treatments like
a human, and is used for training
physicians. On the right is Paul
Clark of the Aerojet-General
Corporation which helped the
university in the development of
the simulator.

Domestic robots

The Japanese inventor Igashichi Iizuka (1762–1836) is best known for his chiming clocks and his attempt at making a flying machine. However, he also made a big doll, bigger than the traditional tea-carrying doll, which he sent on errands, or rather on a specific errand to buy sake. The doll was given a flask and immediately proceeded to travel through an exact distance, making turns at the right places, until it arrived at the shop selling sake. There it waited until the flask was filled, when it would turn round and come back. Should, however, the shopkeeper not pour in enough sake, the doll would not budge, because the mechanism only operated when a certain weight was placed in the flask. Of all the domestic robots to date this doll still appears to be the most functional, because its capabilities are finite and no reprogramming is required for other tasks.

Domestic robots at large have been on the cards since the 1950s and occasionally a date would be given and duly published, when such machines would appear on the market at reasonable prices. Professor Meredith Thring, best known for his work in the area of telechiric robots, ended his lecture at the Royal Society of Arts in 1967 with a statement about 'Robots in the Home': 'Probably more routine work is done by intelligent people in the home than anywhere in modern civilization. All the cleaning, scouring, dusting and daily tidying jobs, preparation of vegetables and so on, are routine, and require no judgement or intelligence on the part of the person doing them. The development of a robot at a reasonable price to act as a slave and do the dull jobs in the home is therefore as worth while an objective as the development of a robot for industry or the farm.'[1]

The domestic robot has become a symbol of progress, as an important innovation which will give people freedom from drudgery. Despite the very amusing and ingenious devices that have been made and are called 'domestic robots', the domestic robot as an idea belongs to that large area of consumer propaganda which has no bearing on reality. Until a robot is as good as man in very many aspects (at which point he will already be better than man in some), domestic work will continue to be done by people. By the time robots do reach this elevated plateau of complexity they may already be considered too good and/or too important for domestic work. The only way out of this 'catch 22' is to begin to think that domestic work is really rather important, that it is not as routine as Professor Thring suggests, and requires the type of intelligence which in a machine will require some more basic discoveries.

Computing and robots often get confused, especially in the minds of children. When, in 1971, a group of London schoolchildren were asked to make a picture of a computer, twenty-five per cent of them depicted a scene from *Doctor Who*,[2] and twenty per cent depicted domestic robots, most of which were armed with vacuum cleaners. How much work the domestic robots did often depended on how much money one fed them with. Apart from vacuuming floors, their tasks included hunting mice, counting and storing food, cooking and pumping water.

When children imagine a domestic robot, they think of a companion in their home; when adults imagine a domestic robot, they think of a servant. In *The Hubbles and the Robot*,[3] a children's book about the future, a schoolboy travels to a historical museum of obsolete things to bring back a Victorian maid of all work. The maid, however, is not exactly a servant although her cleaning is impeccable. She soon gives herself airs, sings sad ballads while accompanying herself on a harmonium and eventually decides to become the mistress of the house. In the adult world, she would be unlikely to be given such an opportunity, even temporarily.

In the adult world the symbols of domesticity are the insignia of office, or the tools which surround the cleaner or of which the cleaner is composed, as is the case with *A Couple Lamenting the Results of Domesticity*, and the two figures by Braccelli, made up of spoons, sieves, bowls, washboard, jugs and scissors. The contemporary domestic symbol is the vacuum cleaner with which most domestic robots are depicted. What is surprising is that the vacuum cleaner with Televox, photographed in 1927, is quite similar in style to that with the Quasar domestic robot made in 1977. The appearance of the robot itself meanwhile has evolved quite considerably, but the tools of office remain the same.

Televox, presented next to the portrait of Westinghouse, receives his instructions by telephone; other robots are operated by remote control, or with a small computer built into their bodies. This is the case with Arok which is

[1] Professor M.W. Thring (Head of Department of Mechanical Engineering, Queen Mary College, University of London). 'Robots' lecture delivered on 4 January 1967. *The Journal of The Royal Society of Arts*, London, April 1967, pp. 387–91
[2] *Doctor Who*, a BBC TV science fiction adventure series, started in 1963
[3] Elaine Horseman. *The Hubbles and the Robot*. Chatto & Windus, London, 1968

LA FEMME DE MESNAGE.

Vous filles Friquettes et Gentilles
Prenes Garde à Vos quoquilles,
Contemples ce pauure Visage:
O due L'embaras du Mariage:

A deguisé en peu de temps.
Malgré tous mes beaux Courtisans
Baste per mon pauure pucelage
Et deuient femme de mesnage

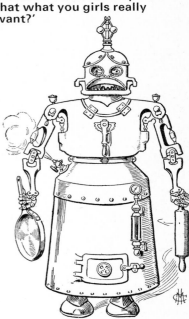

A Couple Lamenting the Results of Domesticity, French, 17th century. The texts give warning to young men and women who might contemplate marriage and the inevitable result is expressed in the two faces overrun by domesticity. 'You give up your virginity to become a housekeeper. Is that what you girls really want?'

A Victorian domestic invention

L.HOMME DE MESNAGE.

Vous garcons Tourlours galans
Charmeurs de gens Tourlours gosins
Quand Nous contemples Set, Vinage
Vous Allez Vn Grand auantage

De Voir que Malgre mes dents
Ie Faw Rire les Regar dants
Par Mon Facecieux Visage
Faict d'ambaras du Mange

Two sets of figures made up of domestic implements, from Braccelli's *Bizzarie di varie figure*, 1624

113

Televox was built by the engineer Wensley of the Westinghouse Electric Corporation in Pittsburgh. It was designed to take orders through an audible code sent to it by telephone, and itself was able to produce a series of sounds which signified that the instructions had been received. It could also pick up the receiver and ask for instructions.

In *Automata*, Alfred Chapuis and Edmond Droz report:

'To achieve this astonishing development, Wensley used sound film. A band of film between 14½ and 18 feet long was made up into a loop. The spoken sentences were encoded onto the edge of the film as visible bands of varying transparencies and widths. A motor set the film in motion and a selector mechanism illuminated the required phrase which was then transmitted to give the illusion that Televox was talking on the telephone.'

controlled by a tiny computer that can be programmed for daily duties such as walking the dog, emptying the rubbish and serving drinks. The repertoire of most domestic robots is very similar although the domestic couple of Claus Scholz of Vienna also answer the telephone, shake hands, introduce themselves and pour liquids from one container to another without spilling anything.

The car-washing robot made by Dennis Weston in Leeds is reputed to be the only robot of this sort to have successfully walked up and down stairs. Called Tinker, it can be programmed to perform 'any reasonable task', although it does take Weston some four hours to program Tinker for the contours of the car, before it can be washed. Made in 1966, it can make 180 separate actions, contains 120 electric motors, a zoom–TV lens, camera eye, a memory, and twenty channels for receiving signals.

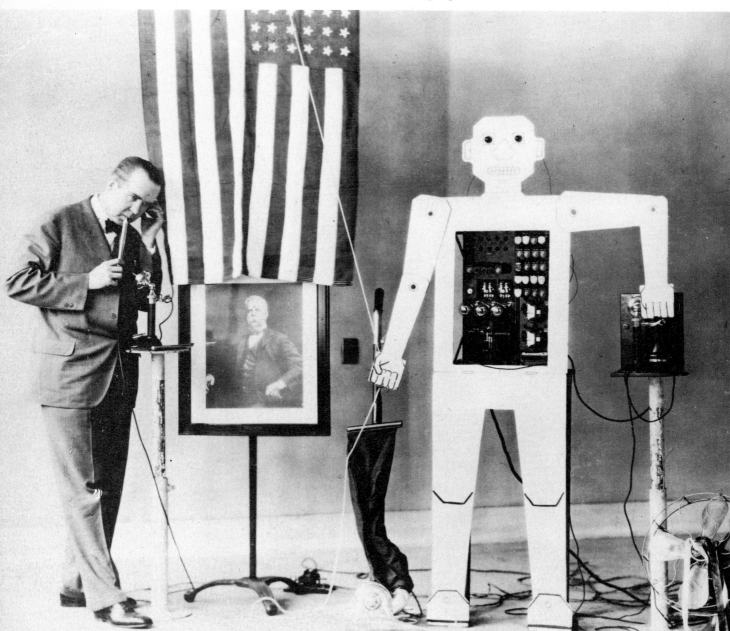

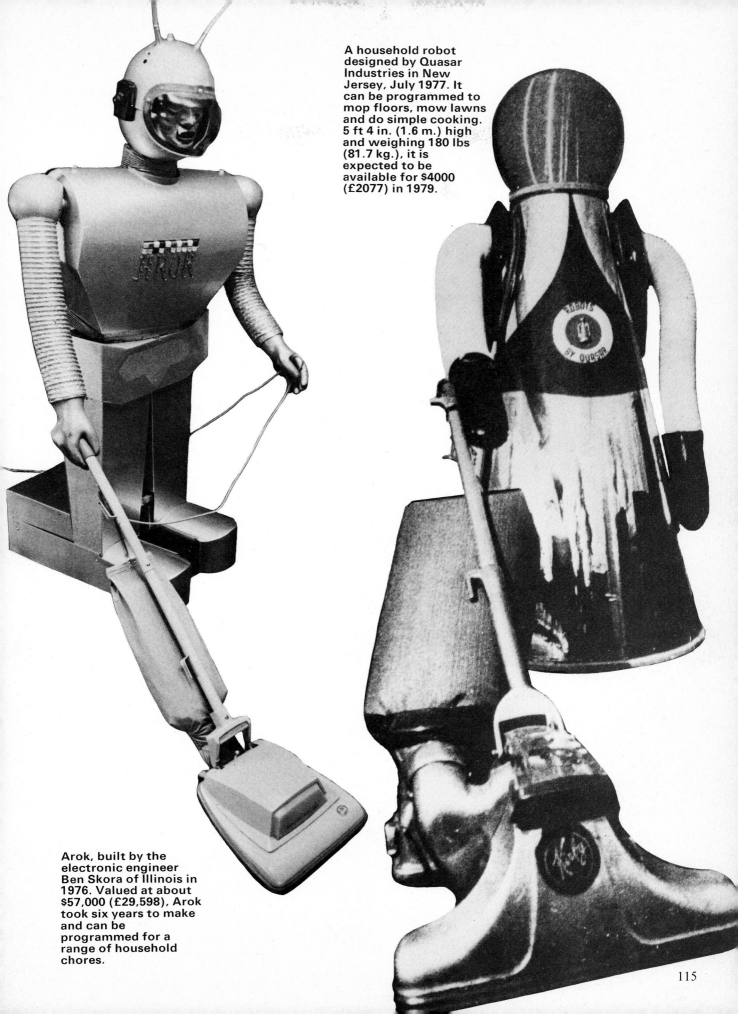

A household robot
designed by Quasar
Industries in New
Jersey, July 1977. It
can be programmed to
mop floors, mow lawns
and do simple cooking.
5 ft 4 in. (1.6 m.) high
and weighing 180 lbs
(81.7 kg.), it is
expected to be
available for $4000
(£2077) in 1979.

Arok, built by the
electronic engineer
Ben Skora of Illinois in
1976. Valued at about
$57,000 (£29,598), Arok
took six years to make
and can be
programmed for a
range of household
chores.

The Reckitt Industrial robot is a computerized cleaner. First presented to the world in January 1977, it is 3 ft 2 in. (0.97 m.) high and weighs under three stone (19 kg.). It can scrub and polish floors, dust and polish desks and furniture, sweep, vacuum, and remove excess water.

Dennis Weston's Tinker, which can be programmed to perform 'any reasonable task'

Military aggressors, defenders, guides and annihilators

There is a story about a battle fought by robots which goes on for months on end. The soldiers of both camps sit in tents in considerable comfort and control the robots by means of a keyboard. With the coming of spring, on one particularly warm day, the robots are suddenly moved by 'the feeling of spring, by love, and everything . . .', and rush out into the green fields full of flowers, where they cavort and pick violets. At this point the soldiers have to get down to the battle themselves. One of the sides soon wins, and on getting home a soldier is heard to say that had the robots not gone suddenly berserk they would all still be on the battlefield.[1]

This story is untypical because the robots obviously indulge in some mild non-competitive game as there appear to be no winners. It was written in France during the war when hope might have seemed to have been better placed in machines than in humans.

Such sentiment must be considered exceptional, because as and when, whether in literature or in the real world, robots are used for fighting wars, protecting property, enforcing law, or taking responsibility for human beings even if it is entirely for their own good, there is an implicit danger. The danger is based on two major factors. The first has to do with the fact that if you have rules, you must have exceptions to those rules, and to date machines have no conception of such an idea. Exceptions are by definition unforeseeable, and there is no numerical or statistical or stochastic system which can enable a programmer to endow a robot with the ability to understand

A specially aggressive robot by Bob de Moor

and implement the necessity for making exceptions. The attribute which is requisite for this is imagination and robots cannot imagine (since they still have no component to give them such a capacity) the consequences of their actions. In this case, predicting results is not the same as imagining results. As a component of robotics, surprisingly enough, imagination is not mentioned to date either in technical or fictional literature, although feelings are. From the point of view of human beings, imagination might be more important as an attribute in such a robot than emotions.

The second, and even more basic factor is biological altruism which must constitute a part of the make-up of a responsible robot, if man is to survive (this is discussed at the end of the book in 'Robots, ourselves, and the future'). Until such time that these two components become obvious issues in robotics, one must make sure that robots used for enforcing law, in any way whatever, are as inefficient as we are, since our salvation often lies in the loopholes of our laws and the inefficiency of their human executors.

The history of the machine gun and its gradual development towards self-government make this point. The three machine guns illustrated here which demonstrate an increasing degree of automation are: a sixteenth-century manual machine gun used for cavalry defence; an 1880s' Maxim self-acting machine gun recommended for defence of battlements and trenches; and finally a US Army 175 mm. self-propelled gun used in conjunction with a computer and an electrical tactical map. The critical moment, already reached with the autopilot tracking missiles, represents the fourth stage when the gun will be autonomous and its communication with the computer will become a closed loop to which man will have no access because such a refinement will have been thought to be no longer necessary. Such a gun in its most simple version could shoot on detecting a disturbance while guarding banks, livestock or goods. Thus, either it must be capable of making exceptions so as not to kill the innocent, or it must be sufficiently inefficient so that it can be outwitted.

There are areas where the usefulness of robots is not in question. Those used in military operations for bomb disposal are invaluable in saving human lives and protecting men from injuries. They are the precursors of robot fire brigades which likewise, under man's direction, can protect human life. Even in traffic a robot can be more effective than either a policeman or road signs or lights. What the traffic robots do is so simple that the mechanism responsible for the action, that of

US Army's 175 mm. self-propelled gun,
with computer, battery display unit and electrical
tactical map

A sixteenth-century manual machine gun, and
the Maxim self-acting machine gun in an etching
by Poyet, 1885

moving an arm up or down, does not even deserve the name of a robot. They are used because motorists are more likely to notice brightly dressed mechanical giants than similarly dressed humans. A humanoid figure has far more chance of attracting our attention than a human figure.

The dividing line between defence and aggression, guidance and misdirection, and even between life and annihilation is so fine that nobody can be absolutely sure to recognize the implications of the technological wonders to which we are introduced. Subtle, but ultimate electronic takeover is discussed by Oswald Wiener in *The Improvement of Central Europe*.[2] He describes in his first sketch for a solution for the world, a device called Bio-Adapter which replaces the world and the entire human environment for anyone who steps into it, producing, at the same time, all possible gratifications mental and physical including the illusion of being in the world. Who should use such a device? Wiener puts forward several categories of people including those who simply want to use it, and those whose services to the state merit a special acknowledgment. Of course, those whose presence is a nuisance could also be encouraged to use it because once in the Bio-Adapter, nobody can get out.

The Bio-Adapter is something akin to a flexible garment which the person to be adapted enters. His–her needs and desires of every sort are studied and provided for, but although the Adapter starts out as an external entity, it ends up as a part of his–her nervous system. The Adapter has precise feedback systems, the latest monitors, it can perform operations, including amputations and neurosurgery, and regenerate its own functioning.

The process by which the machine becomes a part of the man (or vice versa?) is ingenious. At first he–she can walk and eat and converse with the Adapter in human language without being aware that he–she is in an enclosure and not conversing with real people. Later, the activities are simulated within the nervous system. The person meanwhile has the illusion of living in familiar surroundings, those which he–she knows or desires, except that everything is pleasanter than he–she expects. During the second stage, the Bio-Adapter begins to take over the nervous system and gradually gets rid of the body: first the limbs, and then the rest. The human becomes gradually reduced. At this point the Adapter does not have to work so hard because no food needs to be prepared and the person inside has no need or desire for real movement. Then comes the abolishing of the whole nervous system, with the cells being taken over by electronic complexes of the Bio-Adapter. Consciousness is no longer induced by physical means but by constant flow of information produced electronically, with everything gradually becoming consciousness and with the subconscious existing at an atomic level. Finally, consciousness having become self-sufficient creates its own world.

The Bio-Adapter has digested the human. Those outside the Bio-Adapter are physically inadequate, equipped with a miserable nervous system and left to their own devices, while those inside have become an integral part of the cosmos. Millions of these Bio-Adapters are to be found in vast honeycombs under the surface of the earth.

[1] Isabelle Sandy. 'Le Cœur et la machine'. *Le Journal*, March 1940; discussed in Alfred Chapuis. *Les Automates dans les œuvres d'imagination*. Editions du Griffon, Neuchâtel, 1947
[2] Oswald Wiener. *Die Verbesserung von Mitteleuropa* (written 1962–7). Rowohlt, Reinbek bei Hamburg, 1969

A robot bomb detector used by the British Army in Belfast consists of a television camera mounted on a flexible arm surmounting a tank-like base. Video display, safely out of the danger zone, shows what is in the camera's range of vision. The robot assembly, called 'wheelbarrow', also carries a number of devices for diffusing and disarming bombs.

The robot flagging down traffic outside Cologne was introduced to make motorists stop. The experiment must have been successful because a similar robot was placed in Paris near the Arc de Triomphe to warn motorists of roadworks. The notable difference is that the French robot, called Mr Sam, is smiling. Both robots wear fluorescent jackets and are fitted with polyester arms operated by batteries to enable them to rise and fall.

This robot soldier was shown at the New York Exhibition in 1939. It is 9 ft 3 in. (2.8 m.) tall, weighs half a ton, has an 18 horsepower motor and is radio-controlled. It moves on caterpillar feet, with arms stretched out in front ending in wheels bearing a series of fearsome clubs. It is reported that it also contained asphyxiating gases. In the two drawings of 1924, the robot soldiers are busy dispersing terrified crowds, while the guard (apparently the employee of a bank) is sitting in the van with radio-control equipment for guiding the robots.

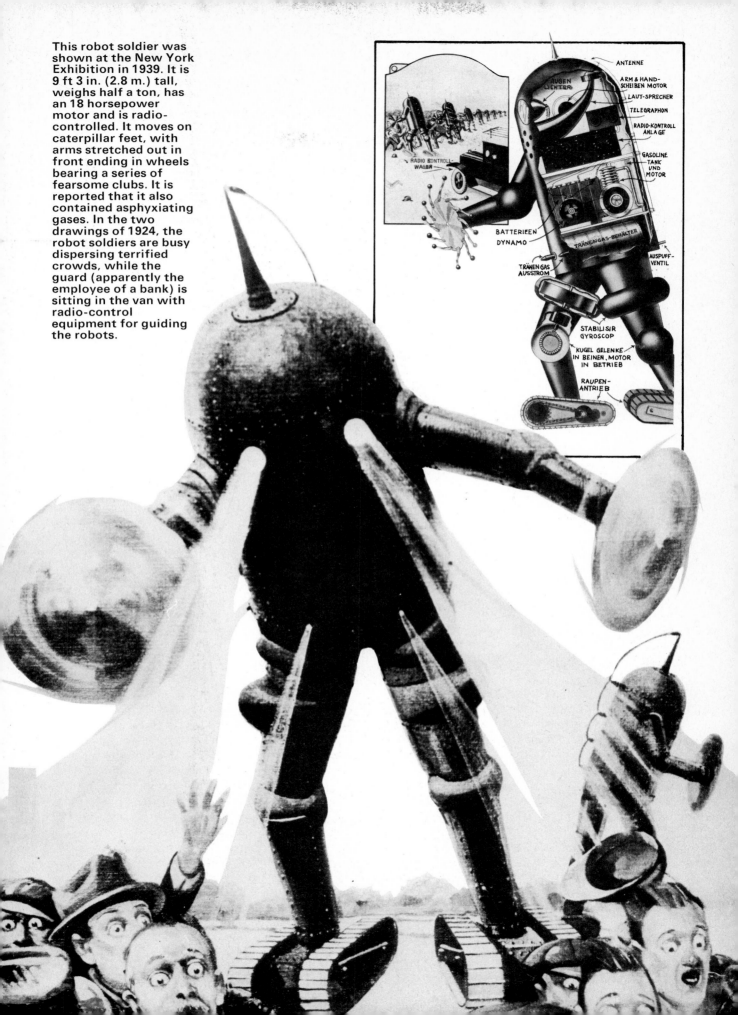

Artificial arms, hands, and other extensions

'Instead of arms and legs they had transparent extensions shining in the sun. Each of these limbs was a tangle of metallic rods and coils with tiny bulbs scattered through which lit up and faded as the limb moved.'

In Bernard Wolfe's *Limbo '90*,[1] a novel about the society of the future, after a great and bloody war men volunteer to have their limbs amputated to protect peace, as it is generally believed that there can be no 'demobilisation without immobilisation'. The premise, which in due course is proved totally false, causes the majority of men under forty to be fitted with new electronic limbs, more effective and efficient than anything made by nature. 'Each new electronic limb was fitted permanently into the stump by cineplastic surgery, connected up with all the muscles and nerves of the stump. Designed so that any kind of limb could be snapped into it and immediately be hooked in with the musculature and the neural system. The movements of the limbs were guided and controlled by neural impulses relayed from the brain through the central nervous system and powered by an atomic energy capsule . . . Electrical energy was translated into mechanical energy.' The limbs were equipped with (among other things) a system of levers and linkages which did the work of the original muscles and tendons but with greater power and control. While a human organism is capable of continuous exertion of not more than one-sixth of a horsepower, the prostheses of *Limbo '90* could reach a level of hundreds of horsepower, since the power came from energy capsules. Possessors of artificial limbs could perform feats inaccessible to ordinary men, and indeed propaganda for enrolling more amputees was based on this fact.

Artificial limbs, such as those described by Wolfe in 1953, still belong to the realm of fiction, although the number of letters received by the Limb Fitting Centre at Roehampton, asking for bionic limbs, suggests that many people who have seen them on television believe that they are already available. Since the goal of making prostheses is the physical, psychological and economic rehabilitation of the amputee, it is unlikely that in the foreseeable future prostheses will be designed which can do more than replace the lost functions of the human limbs.

Even so the progress made during the five hundred years since the advent of the most famous iron hand in history is impressive. The iron hand was made in 1509 for Goetz von Berlichingen, a knight who flourished at the time and was immortalized in Goethe's *Goetz*,[2] a drama in the style of Shakespeare. The hand had gearing for fingers and thumb, but as Goetz himself says, it had limited use: 'My right hand, though not useless in combat, is unresponsive to the grasp of affection. It is one with its mail'd gauntlet – you see, it is iron!' Artificial limbs of the period, as one can see from the designs of the famous military surgeon Ambroise Paré (1510–90), were made in the fashion of armour and were extremely heavy. But they possessed a mechanism for bending the elbow, and a hand that could open and close with fingers kept extended by springs and flexed by ratchets and levers – which made Paré the undisputed precursor of modern prosthesis.

Prosthesis is the addition of some artificial part to the human body (natural replacements taken from other humans are called homografts). In real life a prosthetic device can give an improved function but can never restore completely the normal activities of a limb. So far no substance has been found which could replace muscle and still be controlled by the will. Until such time that the inhibitions of the body to regenerate parts of itself are broken down, artificial limbs will be necessary. The inhibition originates at a certain point in the evolution: a spider will grow a new leg but a dog will not grow a new tail. For humans the solution lies (at least for the present) in symbiotic co-existence with mechanical devices. The closer the artificial limb comes to the patient's own body image the more successful will be the collaboration between the organic and the mechanical parts. The solution lies in the possibility of 'feeding' the phantom limb (an impression that the limb which has been severed is still there) into the artificial arm before the patient gets used to being one-handed.

Even without our limbs, we are more flexible than we realise. When a two-arm amputee returned his artificial limbs to the Limb Fitting Centre at Roehampton[3] because he preferred to do without them, there was considerable surprise. The point he was trying to make was that he could write and eat with his feet, and that the doctors should train our society to allow amputees, like him, to eat with their feet in restaurants and to stop imposing artificial arms and hands on those who would prefer to do without them. However, had this man wanted to play the trumpet, he would have needed to keep his prosthetic arms. Even then things would not be so simple because three fingers are needed to depress the keys of a trumpet and the hook of an artificial arm could only serve

[1] Bernard Wolfe. *Limbo '90*. Secker and Warburg, London, 1953
[2] Johann Wolfgang von Goethe. *Goetz*. Translated by Walter Scott. J. Bell, London, 1799

as two. Ingenuity is obviously more important than arms and legs because a twelve-year-old American boy who had no upper limbs did play the trumpet. He achieved this by using the hooks of both his artificial arms and propping up the bell of the trumpet on his leg.

There are two main types of artificial limbs: unpowered and powered. The unpowered use the movement of the remaining muscles transmitted through a harness, or a system of cables, to activate the limb. For powered upper-limb prostheses, such as those used by the intrepid trumpeter, compressed gas and electricity are the main sources of energy, and a servomechanism is the basis of the control system. Long-life batteries, and miniaturization due to microelectronic chips which contain tens of thousands of components, make it possible for the artificial limbs to be reasonably light. In a typical artificial hand, electric signals from the stump muscles are detected and amplified and provide information for operating and controlling its movements. Electrodes, either implanted in the arm or attached to the surface of the skin, are connected with the leads of an amplifier which actuates the motor. The energy comes from rechargeable batteries. As the electric signals from the stump muscles are proportional to the muscle contraction, the degree of muscle contraction controls the speed of the hand's movement and the force of its grip. When the muscles are relaxed the hand does not move.

A more sophisticated artificial hand, with a gauge which calculates the strain exerted by the gripping mechanism of the hand, produces a tiny electric signal, proportional to the force used, and transmits it to the medial nerve in the arm. The resulting tingling sensation corresponds to the exertion of pressure and enables the patient to control the hand with greater precision.

Powered artificial arms are of critical importance for those who have no arms at all. A person with one healthy arm can probably manage quite well with an unpowered arm which will serve as a prop. The two-arm system, developed in Edinburgh for the victims of the thalidomide tragedy and bilateral amputees, is also based on a pair of arms of which one is powered and the other can simply be moved by being pushed into position.

If an upper-arm amputee retains the original arm muscle still intact, he can be fitted with an artificial limb such as the Boston Arm.[4] He will control his mechanical limb as he would a real one because the messages from the brain are communicated to the remaining muscles and trans-

Description de la main de fer.

1 Pignons feruants à vn chacun doigt, qui font de la piece mefme des doigts, adiouftez & affemblez dedans le dos de la main.
2 Broche de fer qui paffe par le milieu defdits pignons, en laquelle ils tournent.
3 Gafchettes pour tenir ferme vn chacun doigt.
4 Eftoqueaux ou arrefts defdites gafchettes, au milieu defquelles font cheuilles pour arrefter lefdites gafchettes.
5 La grande gafchette pour ouurir les quatre petites gafchettes, qui tiennét les doigts fermez.
6 Le bouton de la queüe de la grande gafchette, lequel fi on poulfe, la main f'ouurira.
7 Le reffort qui eft deffous la grande gafchette, feruant à la faire retourner en fon lieu & tenant la main fermee.
8 Les refforts de chacun doigt, qui ramenent & font ouurir les doigts d'eux mefmes quand ils font fermez.
9 Les lames des doigts.

The iron hand known as 'Le Petit Lorrain', after the artisan working with Ambroise Paré in Paris, was depicted and described by Paré in the middle of the sixteenth century. 'The hand, which on the dorsum has the form of a steel gauntlet, is attached to the forearm by two metal rods and leather straps. The thumb is rigid and the fingers are kept extended by four springs fixed in the palm; when they are flexed, they are kept so by ratchets worked by metal levers.'[5]

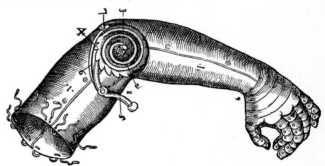

The artificial arm with a mechanism for bending the elbow

mitted from there, in the form of amplified electrical impulses, to the artificial arm. The speed of movement is directly related to voltage. Strain gauges enable the user to be aware of the load he is picking up and a potentiometer at the elbow senses the speed of movement. The arm is based on a simple screw-jack system which raises and lowers it. With such an arm an amputee can raise and move objects of up to 10 lbs (4.5 kg.).

One of the most advanced upper-limb prostheses for patients without arm muscles is the computerized artificial arm from Stanford Research

[3] from conversations with Mr Mirosław Vitali MD, FRCS, Principal Medical Officer, Limb Fitting Centre
[4] Boston Arm, developed by Dr Melvin J. Glimcher, professor of orthopaedics at Harvard University, in collaboration with Professor Robert W. Mann of M.I.T., 1969
[5] *Les Oevvres d'Ambroise Paré*. Gabriel Buon, Paris, 1579

Institute in California.[6] It enables the patient to comb his hair, eat, and scratch his back. It has seven joints and can perform all the movements necessary to accomplish every essential human task. A strap running across the back holds the arm in position and sends signals to a computer which interprets them and actuates the arm. For instance, the elbow lock can be moved by pushing the chin forward; moving the shoulders together causes the arm to move upwards; pushing the shoulders apart returns the arm to the previous position. As many as twenty different co-ordinated movements can be achieved with a single integrated circuit embedded in the arm. The next step is for a pair of co-ordinated artificial hands with fingers able to tie a pair of shoelaces.

Orthosis, which literally means correction of crooked parts, as opposed to their replacement, is also a field for many ingenious solutions. A gas-assisted automatic arm which enables quadriplegics to write is a device strapped on to the patient's inert wrist and hand. The device is triggered by a shoulder movement which opens a valve allowing gas from a portable cylinder to expand a rubber 'muscle' and make the fingers close on to the thumb. Thus the patient can pick up a pen, hold it, and write as he would have before his injury.

The most common orthotic device is a metal stick held between the teeth; it has a single digit terminal at the other end. With this very simple method a patient can type and use a telephone.

Whereas the human arm can only be replaced by something as closely resembling the original as possible, the loss of other functions requires different solutions. A laser cane, as a mobility aid for the blind, is half the price of a guide dog. It is used virtually as an extra limb which becomes an obstacle detector. The Bionic Instruments Inc. laser cane C5 has been in use since 1974.[7] The cane emits a tone of 200 Hz to warn the user of a drop larger than 6 in., about two paces before it is reached, e.g. steps, station platforms and holes in the road. The straight beam ahead will react to obstacles at a distance of 5–12 ft (1.5–3.7 m.), either with a tone of a different pitch or a tactile stimulator which contacts the index finger of the hand carrying the cane. The upward beam detects branches, signs and awnings and produces a high-pitched beep. Since the beams at 10 ft (3.1 m.) are only 1 in. wide, the system allows for precision.

Remote control systems enable quadriplegics to control power devices with those small portions of their bodies which can move. A spectacles frame with a diode embedded in it and a small magnet on the skin near the eyebrow could trigger the mechanism that would open doors, ring bells

or turn pages of a book. In the same way as an artificial arm could be controlled by toe-actuation while wearing sandals specially constructed for the purpose, the clicking or rubbing of teeth could activate a small accelerometer in the spectacles frame which in turn would enable the patient to manipulate things in his environment. One of the most useful systems for those totally paralysed consists of a light worn on a head-band, which activates the key on an electric typewriter when shone on the corresponding letter of the alphabet placed above the patient's head. The movement required is minimal. There is a predetermined delay to allow for shifting from one letter to the next, depending on the speed of the typist.[8]

The bionic man is already with us, but instead of being able to jump the height of a skyscraper or run at 50 mph (80.5 kph) he has come as close as possible to retrieving those normal faculties which he had lost. While machines come to the aid of man to help restore his natural functions, a great deal of research is being done to enable them to acquire human faculties. The experimental arms and hands produced at several universities in Japan are made with dual purpose: to improve prosthetic devices and to study the intricacies of human manipulative skills in order to develop flexible systems of mechanical handling. Dexterity, machines already have, but what the future remote control manipulators will try to acquire is the human hand's sensitivity to texture, temperature and consistency.

[6] The Stanford arm was developed for NASA by Stanford Research Institute, California, in collaboration with Naval Prosthetics Research Laboratory in Oakland, California
[7] The first-generation cane was invented in 1971 by Bionic Instruments Inc., Pennsylvania
[8] The pilot system was invented by D. W. Collins and developed by Hugh Steeper Ltd, Roehampton, 1966

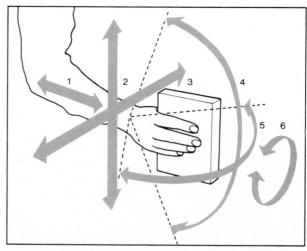

A normal person has about forty-two degrees of freedom in his upper limb but the most sophisticated artificial arm in general use has no more than ten. The diagram shows six degrees of freedom, representing the basic requirements of an artificial hand and arm

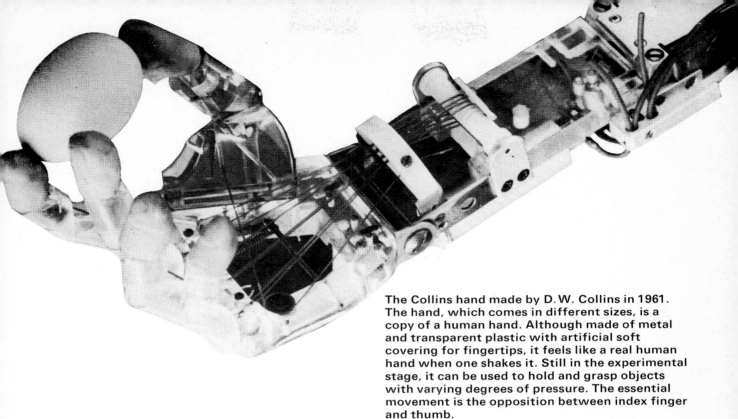

The Collins hand made by D. W. Collins in 1961.
The hand, which comes in different sizes, is a
copy of a human hand. Although made of metal
and transparent plastic with artificial soft
covering for fingertips, it feels like a real human
hand when one shakes it. Still in the experimental
stage, it can be used to hold and grasp objects
with varying degrees of pressure. The essential
movement is the opposition between index finger
and thumb.

▼ The Edinburgh arm produced at the Orthopaedic Bio-Engineering Unit, Edinburgh. Developed between
1963 and 1976, the two-arm system consists of one unpowered arm and one which is powered by gas
and is capable of raising the shoulder and rotating it. When the arm reaches out to grasp an object, the
angle of the hand remains constant so that nothing should be spilled if the hand were holding a glass of
water. The hand is capable of wrist rotation, two types of grip, and has a friction lock shoulder and
elbow joint; it can be adjusted by pressing it against a table or leaning on it. The wrist and the grip of the
hand are powered. The two arms have a total of seven degrees of powered motion.

The early version of the Edinburgh arm was adapted to hydraulic power at the Mechanical Engineering ▼
Department, University College, London, 1975–6. The portable hydraulic power unit is in the left upper
arm. Energy storage in the form of rechargeable batteries is in the waist pack. Change from gas power
to a hydraulic system has an advantage because a gas cylinder, even with a moderate use of the arm,
lasts only a few hours.

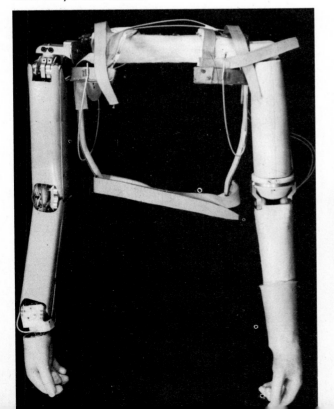

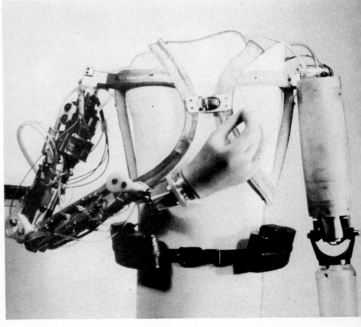

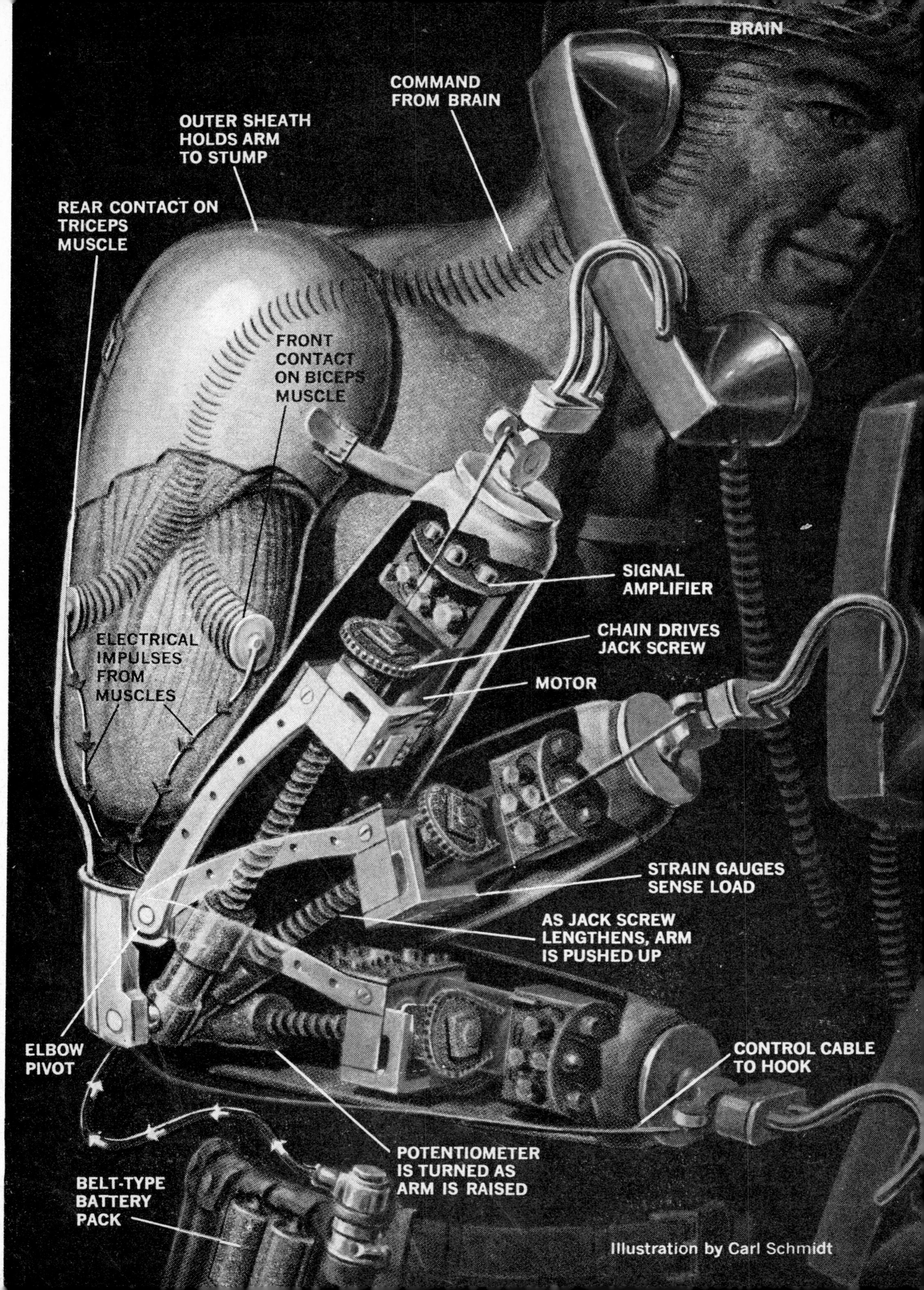

BRAIN

COMMAND
FROM BRAIN

OUTER SHEATH
HOLDS ARM
TO STUMP

REAR CONTACT ON
TRICEPS
MUSCLE

FRONT
CONTACT
ON BICEPS
MUSCLE

SIGNAL
AMPLIFIER

CHAIN DRIVES
JACK SCREW

ELECTRICAL
IMPULSES
FROM
MUSCLES

MOTOR

STRAIN GAUGES
SENSE LOAD

AS JACK SCREW
LENGTHENS, ARM
IS PUSHED UP

ELBOW
PIVOT

CONTROL CABLE
TO HOOK

POTENTIOMETER
IS TURNED AS
ARM IS RAISED

BELT-TYPE
BATTERY
PACK

Illustration by Carl Schmidt

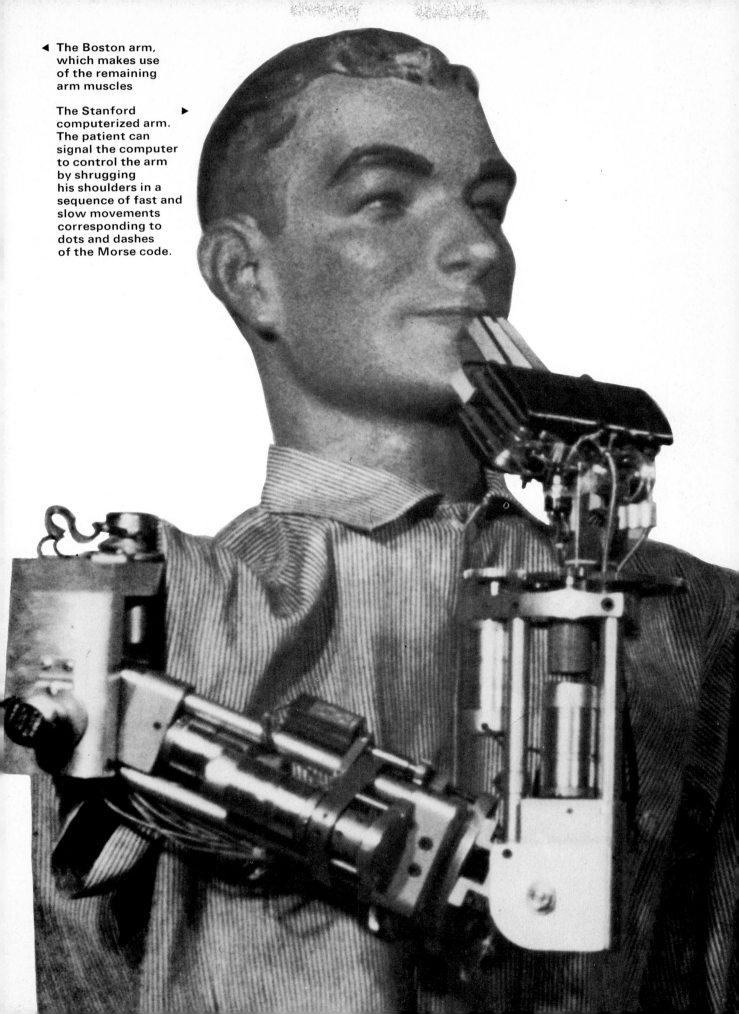

◀ The Boston arm, which makes use of the remaining arm muscles

The Stanford ▶ computerized arm. The patient can signal the computer to control the arm by shrugging his shoulders in a sequence of fast and slow movements corresponding to dots and dashes of the Morse code.

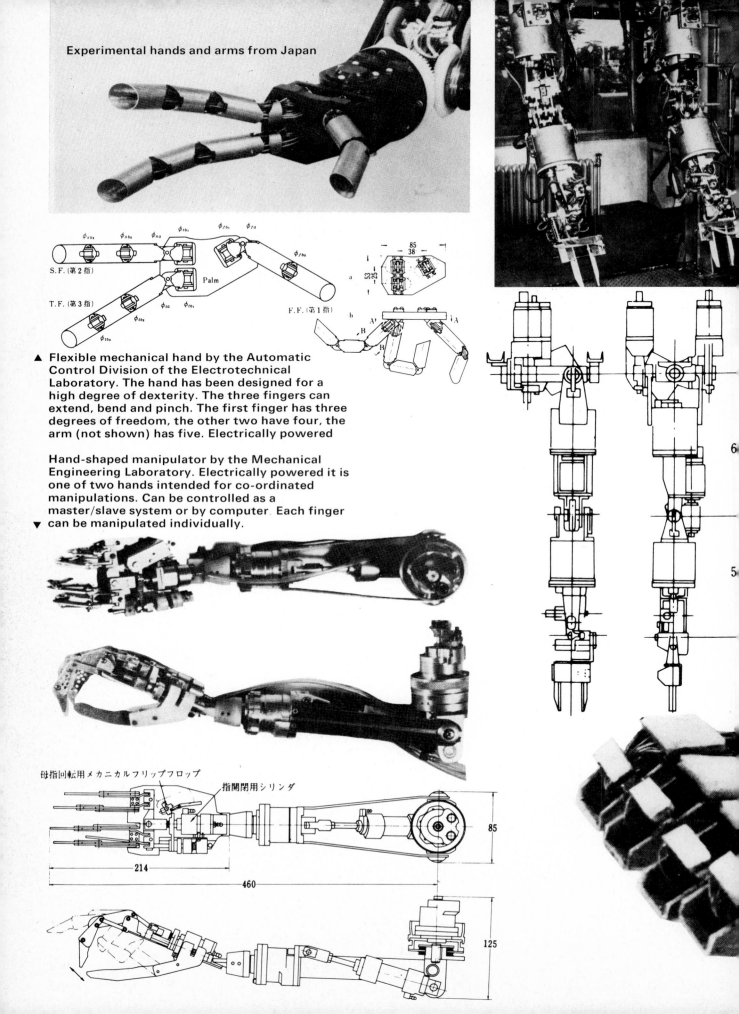

Experimental hands and arms from Japan

S.F. (第2指)

T.F. (第3指)

Palm

F.F. (第1指)

85
38

53
25

a

b A' A
B' B

▲ Flexible mechanical hand by the Automatic
Control Division of the Electrotechnical
Laboratory. The hand has been designed for a
high degree of dexterity. The three fingers can
extend, bend and pinch. The first finger has three
degrees of freedom, the other two have four, the
arm (not shown) has five. Electrically powered

Hand-shaped manipulator by the Mechanical
Engineering Laboratory. Electrically powered it is
one of two hands intended for co-ordinated
manipulations. Can be controlled as a
master/slave system or by computer. Each finger
▼ can be manipulated individually.

母指回転用メカニカルフリップフロップ

指開閉用シリンダ

214

460

85

125

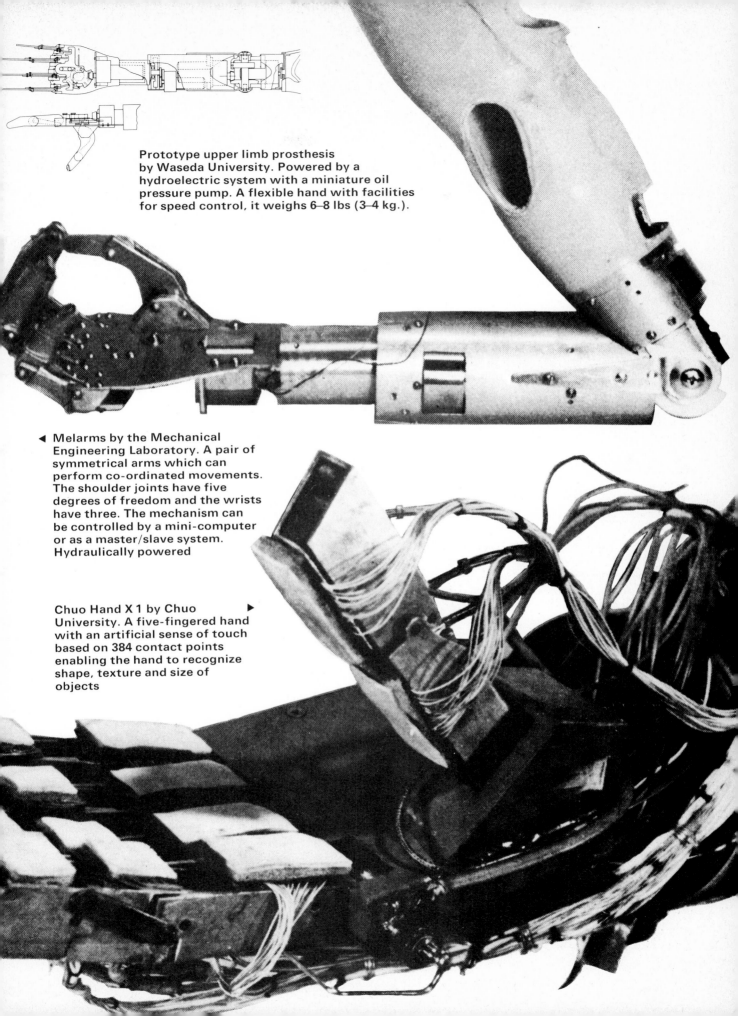

Prototype upper limb prosthesis
by Waseda University. Powered by a
hydroelectric system with a miniature oil
pressure pump. A flexible hand with facilities
for speed control, it weighs 6–8 lbs (3–4 kg.).

◄ Melarms by the Mechanical
Engineering Laboratory. A pair of
symmetrical arms which can
perform co-ordinated movements.
The shoulder joints have five
degrees of freedom and the wrists
have three. The mechanism can
be controlled by a mini-computer
or as a master/slave system.
Hydraulically powered

Chuo Hand X 1 by Chuo ▶
University. A five-fingered hand
with an artificial sense of touch
based on 384 contact points
enabling the hand to recognize
shape, texture and size of
objects

The problems of walking

The first recorded mention of the use of artificial legs goes back to the *Rig-Veda*, at the end of the Veda period in India, 1500–800 BC. Herodotus (485–425 BC) refers to an artificial foot which the seer, Hegisistratus of Elis, made for himself as, when imprisoned by the Spartans, he had to cut off a part of his own foot to free it from the stocks. There is evidence of amputations and the use of artificial legs in pre-Columbian America; they are also referred to in the Talmud and in the Nordic sagas. The Roman mosaic at Lescar in the Pyrenees in France depicts a man with a peg leg which would have been made with a piece of thin bronze fastened with bronze nails to a wooden core and lined with leather. A square piece of iron at the foot is thought to have been added for strength. While early history is full of references to artificial legs, there is little mention of artificial arms. Arms, though they can help each other, are complete in themselves; a single leg is helpless without its companion or a crutch. There are at least seven cases of some sort of crutch for every artificial arm.

Ambroise Paré's artificial legs had a moveable knee and tarsal section of the foot, and a knee lock. Made in the fashion of armour and often elaborately decorated, they would conceal mutilations suffered in battle and were generally too heavy to wear except on horseback. His wooden legs for the poor were far more practical. The stump fits into a hollow wooden bowl surmounting the wooden peg, not unlike the wooden legs in Bruegel's painting of a group of beggars of 1568. The basic design of a wooden socket and a shank has remained the same, though with some improvements, until the present. Two hundred years after Paré, artificial legs had gained considerably in flexibility, although judging by the description of one of the most celebrated legs at the time, it made a noise. The leg made for the Marquis of Anglesey had a steel knee joint and a wooden ankle joint, with cords from the knee to control the ankle motion. It was called a 'clapper leg' because walking was accompanied by a clapping sound. Having undergone many improvements it became known as the American leg.

Unlike artificial arms, which are sometimes powered by electricity, gas or a hydraulic system, all artificial legs are body-powered. An artificial leg for a uniped will never enhance his performance beyond that of walking with two good legs. At best it can allow him to walk without pain, without noise and without frequent need for servicing. Also, the new leg should feel nice and not be too heavy.

However, locomotion with one leg also has certain possibilities. An experiment with a one-legged robot has recently been tried in Japan, as an attempt to produce a system which will be as stable as a two-legged human, whether static or in motion. Their one-legged jumping robot can stand upright, and climb up and down slopes or steps by jumping without falling over, even if the ground on which it lands is unstable. Produced by the Mechanical Engineering Department of Tokyo University, it is the first of a series of

Franciszka Themerson
Bayamus, 1949

experiments dealing with balance relative to speed and motion in general. If a lower-limb amputee were given a chance to consider such an alternative as a single jumping leg, once it had been satisfactorily developed, the very notion of walking could change radically for the few who would be willing to experiment.

In literature the notion of walking as we know it is challenged by Stefan Themerson in his book *Bayamus*,[1] whose central character of that name possesses an extra leg. His system of locomotion, superior to that of bipeds, is to propel himself with great speed on a rollerskate attached to the middle foot while striking the ground with the other two, thus reaching the speed of 50 mph (80.5 kph). He maintains: (1) that his central leg is not a more extraordinary phenomenon than the single leg of a person who has lost a leg as a result of some political activity; (2) that he is not a freak but a mutant; (3) that his mission is to begin a new variation of tripeds; (4) that now, as the earth becomes smoother, Nature is at last free to develop a biological rollerskate under the sole of man's foot; (5) that the main, and quite novel, problem for Nature would be to produce a wheel: a disc capable of rotating about its own axle, but that one cannot imagine how it could organically develop under the sole of a human foot.

[1] Stefan Themerson. *Bayamus*. Editions Poetry London, 1949

A Roman with a peg leg. From a mosaic at Lescar, Pyrenees, France

Ambroise Paré. Artificial legs intended
▼ principally for knights on horseback

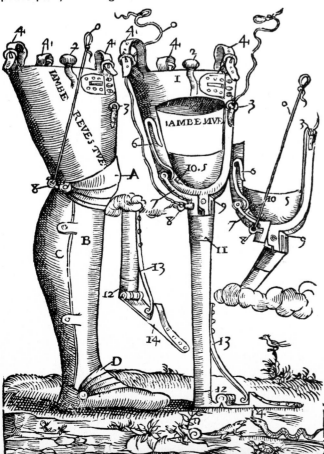

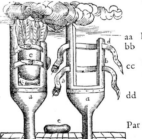

Description de la figure de la iambe de bois pour les pauures.

aa   Represente l'arbre de la iambe.
bb   Les deux fourchons pour inferer la cuiffe, dõt le plus court fe doit mettre dedãs iãbe.
cc   Te monftre le couffinet lequel fe met pour fupporter mollement le genoil fur la rondeur de l'arbre.
dd   Sont les courroyes auec boucles trauerfantes en deux endroits, les fourchons de la cuiffe pour la ferrer & tenir entre iceux.
Par e. t'eft marquee la cuiffe, à fin de t'enfeigner la vraye pofition d'icelle fur ladicte iambe de bois.

Artificial legs for the poor

Artificial limbs by F. Lacroix et Fils, Paris, 1915

131

Pieter Bruegel the Elder. *The Cripples*, 1568. Louvre, Paris

Although the human body could benefit if Nature gave it a wheel, at present it is machines that copy Nature rather than the other way round. With walking robots, as with prosthetic devices, the greater the anthropomorphism the greater the efficiency and economy in consumption of energy, although to make an exact copy of the human leg would be extremely difficult since it consists of twenty-two joints, has thirty degrees of freedom and seventy-one muscle drives. Despite a tendency towards anthropomorphism, a two-legged vehicle, although useful, does not have the stability of one with four legs or, even better, with six.

Walking tractors and mechanical horses have been used in farming since the 1940s. In the Soviet Union, stepping excavators clear forests in Siberia. On uneven terrain, walking machines can be more useful than vehicles on wheels or caterpillar tractors, although they require comparatively more energy for operation. Inertia is more difficult to overcome with the increase of speed but the speed is limited anyway because the machine is not designed to run, and its manoeuvrability has its bounds. Balance is once more a problem: curiously enough the walking lorry can more easily keep its balance when in motion than when stationary. Artobolevskii[2] describes a four-legged vehicle called 'Mechanized horse' which weighs $1\frac{1}{2}$ tons (1.5 tonnes), is 10 ft (3 m.) long and has a 90 horsepower automobile engine. Each leg consists of three sections, all moving in one plane. It can take a $\frac{1}{2}$ ton (500 kg.) load and move at 6 mph (10 kph). Speed and length of stride are manipulated by levers and pedals. He also mentions a sort of train used for military transport. It consists of a series of carriages on legs made of sections which can either push or pull; it can travel over uneven soil, approach obstacles at forty-five degrees when fully laden, and descend at the same angle. It can move through mud, sand and gravel and pass between trees which are as little as 51 in. (130 cm.) apart.

The General Electric four-legged lorry built for the US Army has similar capabilities. The machine mimics the movements of its operator; thus its right and left front leg are manipulated by the operator's right and left arm respectively, and its right and left rear leg by the operator's right and left leg. A six-legged automatic walker could perform better than one with four legs, but one with eight legs is found to be somewhat surprisingly less efficient. The walking lorry has tactile sensors and position-sensors in leg joints as well as a distance-measuring device for scanning the environment in front of it.

132

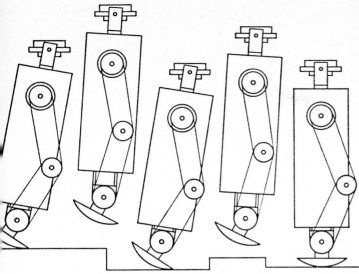

One-legged robot developed at the Department of Mechanical Engineering, Tokyo University

управляющими движениями оператора и движениями ног машины, между моментом, когда сопротивления прикладываются к ее ногам и их подвижным сочленениям, и моментом, когда эти сопротивления чувствует оператор. Оператор сидит и поэтому испытывает при движении машины значительно меньшие усилия, чем их испытывает человек при ходьбе, когда воспринимает

A mechanized horse and a walking train by V. Kovinev

Walking machines have also been constructed for medical purposes. A human-size three-legged walker was built at the University of Wisconsin to study the problems of locomotion, stability and control. It is powered by compressed air, and can carry heavy weights placed high above the ground. The length of each leg alters as the robot is in motion, and the central leg is the principal pusher. The walking robot was developed in the course of research towards the construction of an exoskeleton which would enable paraplegics to walk. Powered by an electric motor, under computer control, the patient is ensconced inside the shell, which can walk, sit, stand up, step over obstacles, and climb the stairs. An awkward alternative to the wheelchair, it may one day provide a walking harness that is less bulky than the present experimental model and less confining than the wheelchair.

Mechanical muscles for paraplegics are intended to restore their lost ability to move. Other mechanical muscles are intended to amplify the ordinary unimpaired human force to a level at which a man can lift a 1500 lbs (680 kg.) load without effort. One of the best-known exoskeleton prototypes, Hardiman, was built by General Electric in 1966 for warehouse and factory operations, bomb loading, underwater salvage, erecting shelters on unprepared sites and in remote places where it is not possible to have other help.

The devices for paraplegics will work only well below human capacity, but the purpose of Hardiman is to exceed the norm. The machine for paraplegics is based on a system in which the patient instructs the computer through a control box; Hardiman is based on a system of negative feedback which goes back to the wearer. The man is the decision maker and consequently, as his movements and force are sensed and amplified, he must be fully aware not only of his own movements but also of the force exerted by the machine. The exoskeleton can be dangerous because it is unlikely to be able to tell the difference between the man's intentional and involuntary movements. The man inside it, misjudging the weight of a load, can over-react and tear a muscle. Once something goes wrong, even very slightly, with the man in control, a chain of actions can develop which the machine will not recognize as mistakes requiring the alarm to be raised and the machine to switch itself off. The symbiosis where a considerable part of man's body surface interacts with a machine is still a very uneasy one.

[2] Ivan Ivanovich Artobolevskii and Aron Yefimovich Kobrinskii. *Meet the Robots.* Molodaya Gvardiya, Moscow, 1977

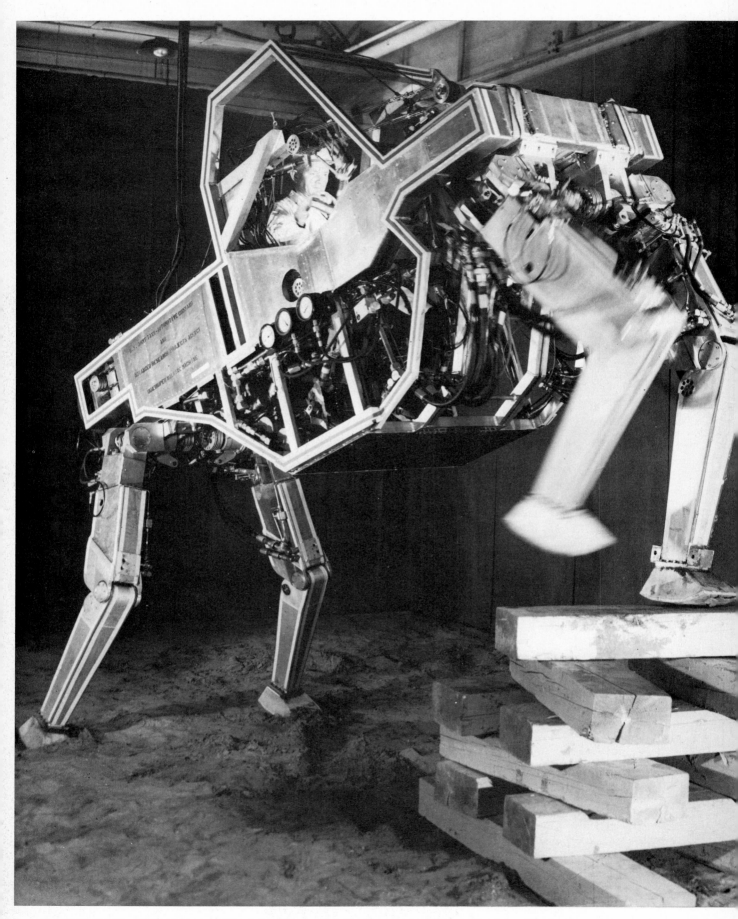

Four-legged lorry by General Electric, Cybernetic Anthropomorphous Machine Systems

Diagram of a three-legged
walking machine and a
photograph of the machine,
developed at the Department of
Mechanical Engineering,
University of Wisconsin,
Madison, 1974

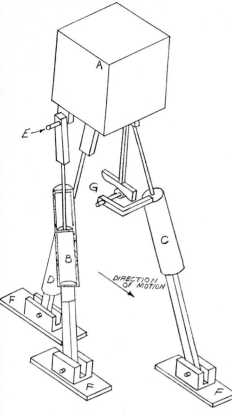

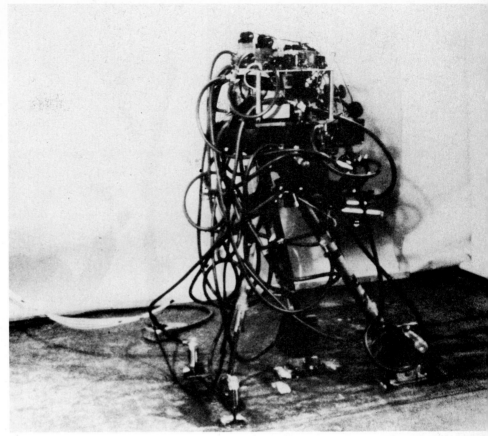

Walking shell for paraplegics, ▶
consisting of 60 lbs (27.2 kg.) of
motorized, battery-powered
tubing and hydraulic joint
actuators. The exoskeleton can
walk with or without a human
load, move forward, backward,
turn and walk up stairs.
Developed at the Department of
Mechanical Engineering,
University of Wisconsin,
Madison, 1976

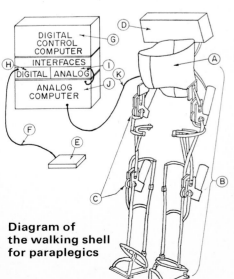

Diagram of
the walking shell
for paraplegics

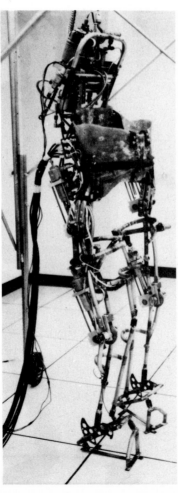

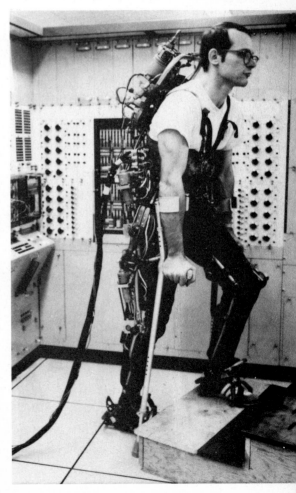

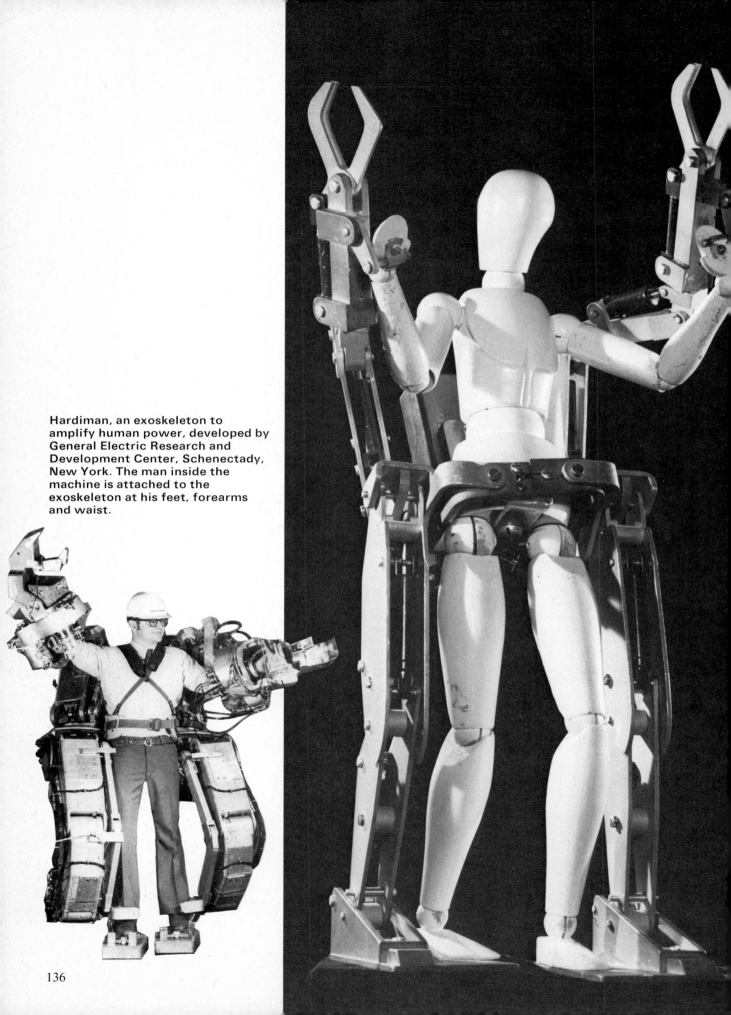

Hardiman, an exoskeleton to amplify human power, developed by General Electric Research and Development Center, Schenectady, New York. The man inside the machine is attached to the exoskeleton at his feet, forearms and waist.

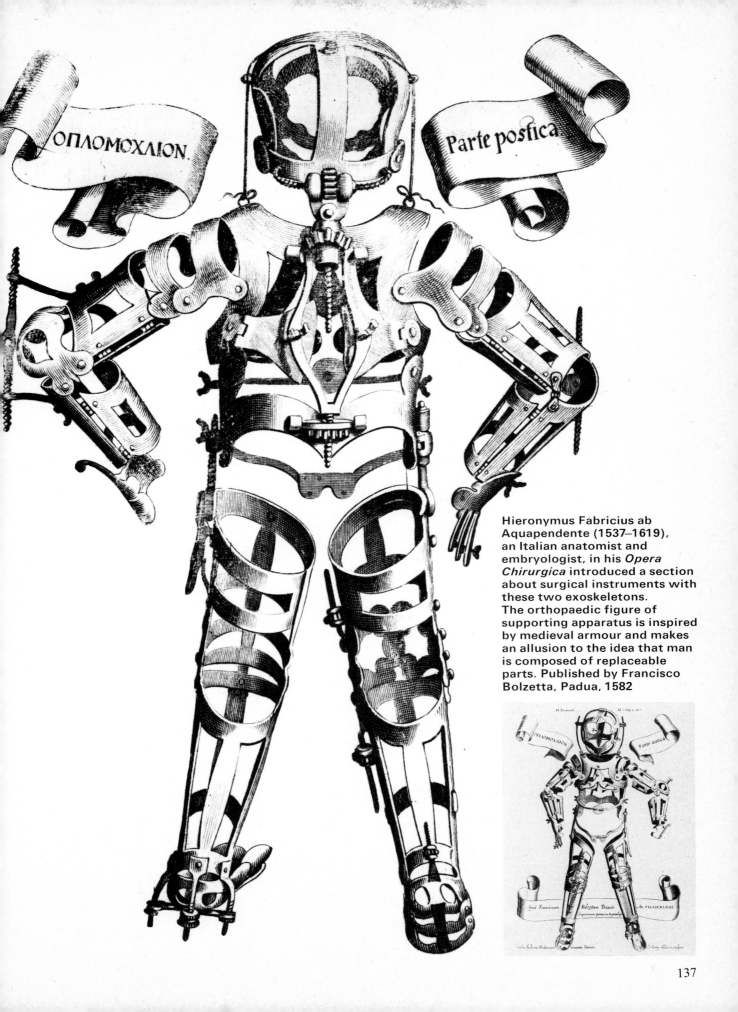

Hieronymus Fabricius ab
Aquapendente (1537–1619),
an Italian anatomist and
embryologist, in his *Opera
Chirurgica* introduced a section
about surgical instruments with
these two exoskeletons.
The orthopaedic figure of
supporting apparatus is inspired
by medieval armour and makes
an allusion to the idea that man
is composed of replaceable
parts. Published by Francisco
Bolzetta, Padua, 1582

To work! To work!

'In the year 1827, a Committee of the House of Commons was appointed to examine into the subject of Emigration – that is, to see whether it was desirable and practicable to remove distressed labourers from the United Kingdom to distant places, where their labour might be profitably employed to themselves and others.'[1]

This drastic measure was contemplated as an antidote to the mechanization of industrial operations and the subsequent loss of jobs.

One of the men interviewed by the Committee was a hand-loom weaver, who explained that working from eighteen to nineteen hours a day, he and his colleagues now earned between four and seven shillings a week, whereas twenty years before, i.e. before the introduction of power looms, the hand weavers could earn a pound a week. The drop in his earnings was attributable directly to the introduction of machinery, but he admitted that machinery neither could nor should be stopped. Others, on the contrary, believed that if they sabotaged machines, mechanical progress could be prevented. What the report attempted to do was to explain to the workers why machines are necessary and why the increased production towards which they lead will in the end benefit them to a greater extent than they can at present imagine. With a steam engine, it was explained, one can do the work of twenty men for fifteen pence, while men earning fifteen pence a day would cost twenty-four shillings. Examples of the advantages of machinery over man-power included travel, coal mining, steel production, road building, and irrigation.

What the report failed to explain to the workers was how they were to survive during the period between the advent of the machine and the consequent growth in productivity which would bring with it new employment.

J. F. Engelberger, President of Unimation, Inc., reports: 'In 1967 the Department of Defense completed a research project labelled *Project Hindsight*. The researchers concluded that an innovation will gain acceptance only when there is a conjunction of three elements: (1) a recognized need, (2) competent people with relevant technological ideas, and (3) financial support.' He concluded that the time of the industrial robot, as an idea whose time has come, is NOW.[2]

Now, as before, human labour will be used as long as it is cheaper. The advent of robots is an economic issue which will continue to bring with it both enormous advantages and enormous problems.

.   .   .

Industrial robots can only work in those factories which have been prepared for them. The disorder in a factory, which any human can cope with, is something that a robot cannot manage. Any human can sort out different components jumbled up in a box and place them in a given order, with the correct orientation, on a work bench. But a robot can only be effective once this has already been done. Even though rudimentary sight and tactile sensors will enable a factory robot to sort nuts from bolts, it may still be economically necessary to program the robot's work as a sequence of actions rather than goal-orientated behaviour. One has to prepare and present data in a way in which a robot can use them, which means that the cost of equipping a factory for robot operation may be ten times the cost of the robots themselves. The need exists, therefore, to design robots capable of working in a moderate disorder, with some ability to recognize colours, shadows, markings, and textures. Experimental robots of this kind are being developed in Japan, the USA, and in a few British universities.

[1] The Society for the Diffusion of Useful Knowledge. *The Working-Man's Companion. The Results of Machinery, namely Cheap Production and Increased Employment, Exhibited: being an Address to the Working-Men of the United Kingdom.* Charles Knight, London, 1831

[2] J. F. Engelberger. 'Economic and Sociological Impact of Industrial Robots'. American Nuclear Society, Proceedings of the Conference on Remote Systems Technology, Hinsdale, Ill., October, 1971

Electro-hydraulic manipulator under development at the Department of Mechanical Engineering, University College, London. A suitable interim measure between a manned and an unmanned factory. The arm is distinguished by a combination of a good dynamic performance with high power and small size.

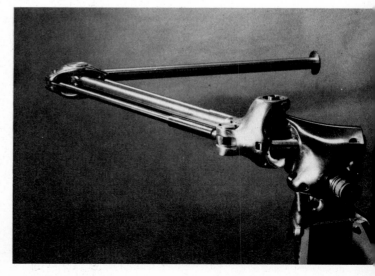

Three Auto-Place mini-robots are used in combination to transfer sheet metal parts between three stages of a stamping press. The entire process takes four seconds. In the stamping sequence a flat sheet is fed from a magazine loaded by the operator. The sequence occurs automatically until such time that the air sensors detect that the magazine is empty.

▲

Man-Mate Industrial Manipulator designed for work in a forge shop. The operator sits in an insulated air-conditioned cab manipulating the boom arm which can reach up to 24 ft (7.3 m.), and can work in temperatures up to 135°F. The system is based on a servomechanism with force feedback, which means that the operator can sense the power which is being exerted through the control mechanism.

▼

Essentially the work consists in matching the manipulative skill of a robotic arm with a computer concept of the real world acquired through some form of artificial perception such as television, sonar, or touch sensors. In the majority of such systems the role of the manipulator is purely executive, but in the case of the arm at the Department of Mechanical Engineering, UCL, the arm itself is designed to be the principal means of obtaining information about such characteristics of objects as mass, friction, and stiffness. Such a manipulator could provide a transition between a manned and an unmanned factory.

Several industrial robots already have visual sensing ability but they represent exceptions rather than the rule. A Mitsubishi robot, for instance, uses two television cameras, one mounted on the robot's hand and the other overlooking the work-bench. The idea is to match the images. When the two images match, the robot hand has reached the right object on the bench. A Hitachi robot, meanwhile, has a sense of touch and can insert a piston into a cylinder with a clearance of twenty microns in three seconds.

The majority of industrial robots are not as sophisticated but can still perform an enormous variety of tasks. Robots can service machines such as lathes, grinders, forging presses, stamping presses and machines for die-casting, injection-moulding and drilling. They can insert pins and rods, mesh objects together and perform some assembly tasks. They can do spot-, arc-, and seam-welding. They cut, pierce, polish, and paint. They carry and transfer objects that are heavy, hot, cold and dangerous, and perform tasks which, for men, would be monotonous and debilitating.

Robots outside the industrial context fight fires, operate in environments in which man would not survive, deactivate bombs, drill underwater, and service orbiting satellites.

To see robots in perspective, they must be considered in the context of automation as a whole.

Robots in automation

The term automation has been in use only since 1947. It means mechanization of operation. There are nine distinguishable stages, or degrees, of automation in manufacture,[3] which can be demonstrated with the example of bending a pipe. The robot enters the process of automation at the fifth stage.

0. Man bends a pipe with his bare hands.

1. The pipe is bent with a hand tool.

2. The pipe is bent with a powered tool.

3. The pipe is bent by powered machinery under human control. (Man-Mate Industrial Manipulator, by General Electric, comes in this category.)

4. The pipe is bent by powered machinery automatically executing a sequence of operations but designed from the outset to perform this, and no other task.

5. The machine bending the pipe is capable of performing a number of operations which can be specified and given a sequence by some program controller such as a patch panel (a control panel on which all the operations can be preset). The program however can specify only the sequence in which control devices, such as switches or valves, are to be turned on or off, and sometimes the length of time between operations. The control of any movement is achieved by preset mechanical stops (Pick-and-Place, or Point-to-Point robots, such as Auto-Place robots).

6. At this stage of automation several programs are stored in memory devices and can be selected automatically. Machines with these characteristics are called variable sequence robots.

7. The degree of automation involves programs stored in larger memory devices and which are changed by signals generated by other mechanisms. On-line changes are accommodated automatically. Continuous-path robots, employing servomechanisms, belong here. In these machines the path to be followed by the tool in such operations as welding, spraying, or profiling, can be specified precisely by means of a computer-generated program, or memorized during a 'training session' when the tool is guided by hand, and the process can be subsequently repeated any number of times.

8. Now the pipe-bending station is under complete control of a computer-aided manufacturing system. The N.C. (numerically controlled) robots are metal-working machines in which the controls, such as levers and hand-wheels, have been replaced by motors directed by programs stored on punched paper tape.

9. This stage, at which we are about to arrive, employs the so-called 'blue collar' robots with tactile and visual capabilities which replace humans on a manufacturing line. They are

sometimes described as 'Intelligent' robots, which, or who, can decide their behaviour by themselves through their sensing and recognizing capacity.

There are two definitions of a robot. The first is used by those working in industry and the second by those involved with research:

1. programmable manipulator of versatile automation components
2. artificial intelligence machine with human-like functions.

Robot types

There are four ways of classifying industrial robots.

1. Degrees of freedom
   This generally means the number of independent motions that a machine can perform. It can be shown mathematically that any manipulator requires three degrees of freedom to position an object in space and a further three degrees of freedom to orientate it in any direction. Thus a general-purpose manipulator requires a minimum of six degrees of freedom.

2. Method of articulation
   This refers to the type of joints employed: the articulation is called *polar* (ASEA) if the joints swivel; *cartesian* if the joints slide (Mr Aros); and *cylindrical* if vertical- and reach-motions are cartesian, and sideways-motions are polar (Versatran). (Wrist motions are always polar.)

3. Control of motion
   Point-to-point and continuous-path. In point-to-point control, only the terminal point of motion is specified, while a continuous-path robot can control also the precise path and the velocity of the entire movement.

4. Method of actuation
   Electric, hydraulic, or pneumatic. Electric motors are popular because they are simple to install and easy to maintain, but their dynamic performance and power to weight ratio is inferior to that produced by hydraulics. Both exhibit good positional accuracy. This cannot be said of pneumatics, which are cheap and easy but suitable only for point-to-point operations.

It is predicted that the robot will come into its own within the next decade. By then it will have multiple appendages and at least two arms, with bilateral co-ordination. It will be smaller, have more energy using less power, and will respond to commands from the human voice. It will be equipped, as a matter of course, with a vision system and tactile sensors.

A committee of the Japan Industrial Robot Association points out that 'the industrial robot will have functions not only similar to those of humans, but also those of animals, such as snakes and other living creatures.' (see p. 147)

A robot is not likely to become redundant because its flexibility has many advantages over purpose-built machines. When a job performed by a robot becomes obsolete, it can be re-programmed to do something else. A prototype unmanned factory, manufacturing machine-tool components, will go into operation in the early 1980s in Japan. One hundred years hence, the factory may still be there, and the robots may still be there, but instead of machine-tool components, they might be producing cakes.

Work at a distance

One of the unquestioned prerequisites of an industrial robot is reliability. It is a thousandfold still more crucial in remote manipulators used for servicing satellites in orbit, handling cargo in space shuttles, assembly of large structures in space, unmanned surface exploration of planets, sample analysis in sealed space laboratories, deep sea exploration, operations in nuclear reactors and other hazardous laboratory environments.

The control of remote manipulators is usually shared jointly by man and computer. Dr Antal K. Bejczy of the Jet Propulsion Laboratory stresses that the best formula in man/machine col-

A Unimate demonstrating how it works

[3] 'Special Report No. 682 on Robots in Metalworking'. *American Machinist*, New York, November 1975

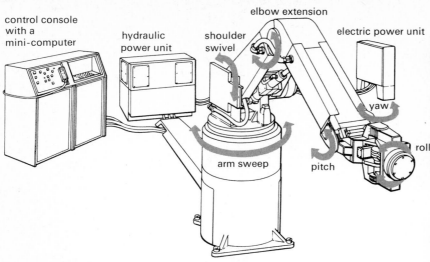

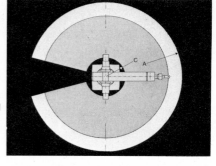

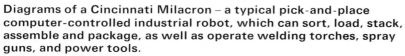

Diagrams of a Cincinnati Milacron – a typical pick-and-place computer-controlled industrial robot, which can sort, load, stack, assemble and package, as well as operate welding torches, spray guns, and power tools.
A jointed-arm robot, it can be remotely controlled and placed in any attitude.

▼ ASEA robot from Sweden. The robot can weld, grind, polish, and paint. The movements of the robot are programmed with the aid of a portable programming unit. Step by step, each position is stored in a memory, and a single programming unit can be used for several robots. Electrically powered, it is designed to handle up to 132 lbs (60 kg.). Four such machines can be supervised by one operator.

Robot with handling capacity	A	B	C	D	E	F	G
6 kg	1159	670	289	200	1620	1150	414
60 kg	2288	1280	989	400	2150	1600	0

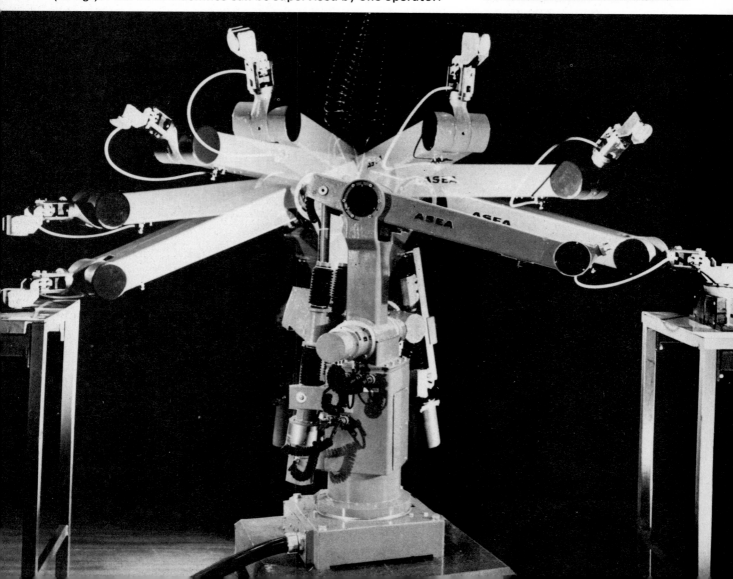

laboration is to use to the full those characteristics in which each excels. Machines are good at continuous monitoring of input and output but they are not flexible enough to deal with contingencies. Man is good at dealing with context and symptoms and can perform tasks without being given complete information, at which point the machine simply stops. 'Human supervisors are used in conjunction with robots because the two are complementary: we automate what we understand and can predict and we hope the human supervisor will take care of what we don't understand and cannot predict.'[4]

One of the most difficult briefs for a remote manipulator of quite exceptional sophistication would be: 'pick-up the (heaviest) object in the (most arbitrary) location and orientation, move it (in the least time) and make the (tightest tolerance) assembly to another object (whose position is poorly known)'. Since the majority of remote-handling robots operate within a less sophisticated range of demands, such a brief would be beyond them; to produce more versatile specimens would be uneconomical for most purposes.

The sort of intelligence a robot must possess for many tasks needed in research is demonstrated by the following hypothetical exercise. The robot must find its way to a given destination using a city bus service. Suppose a number of bus lines lead to a number of places close to the desired

[4] T.B. Sheridan. 'Modelling Supervisory Control of Robots'. *On Theory and Practice of Robots and Manipulators*, preprints of second CISM-IFT oMM International Symposium, Warsaw, 1976. Polish Scientific Publishers, Warsaw

Two views of the AMF Versatran cylindrical robot, capable of the following motions: horizontal, vertical, swing, traverse. The end of the arm can rotate, sweep and yaw. Versatran is used in automotive, die casting, textile and plastics industries, for loading and unloading machines and conveyors, spraying, spot-welding, and palletizing. It can handle yarn, paper, steel castings, plastics, glass, sand and clay.

Mr Aros, a Hitachi arc-welding robot with a miniature non-contact sensor. The sensor allows it to compensate for the inevitable movement of the work-piece due to the heat generated during the operation. Driven by an electro-hydraulic servomechanism, the robot is controlled by a teaching system with instructions stored on cassette tapes.

destination; without knowing which bus will appear first, the robot has to cope with all alternatives in order to plan what to do next.[5]

Remote control, such as that for the Jet Propulsion Laboratory slave arm (which is a good example of the sort of intelligence a robot must possess), can be effected in several ways: with a master arm worn by the operator, a manual control console, and a mini-computer. The same principle applies whether the slave arm is in the room next door or in Space. If it is in Space, then – because of the great distance – there is a time delay between a task performed and the new instruction being issued, and consequently the manipulator is programmed to 'move-and-wait'. The JPL slave arm has eight degrees of freedom, and four proximity sensors (these are small electro-optical devices which can focus on anything a few centimetres in front). The arm has a humanoid hand with three sensors. The sensors are connected to loudspeakers and the pitch of the tone generated by the voltage output of the sensor gives information about the distance between the sensor and the object. The operator can listen to the tones generated, watch television displays, and act on this basis.

Among the most impressive feats in remote control manipulation to date are the achievements in space exploration, such as the NASA Viking 2 spacecraft which landed on Mars in September 1976. It conveyed pictures of the surface of the planet back to earth, collected samples of soil from beneath a rock, having first pushed it aside, and transported them to the automatic organic chemistry laboratory for analysis. Another great achievement is the Soviet robot Progress 1, the first automatic transport spaceship to take fuel and scientific equipment to the Salyut-6 plug-in space station, in January 1978. Unmanned space vehicles, such as this, will enable astronauts to stay in orbit for very long periods at a time.

At the other extreme, on the ocean floor, remote manipulators are used for exploration, salvage, construction, and rescue missions. An underwater robot arm can perform the same tasks as a diver: opening and closing valves, drilling, cable cutting, and filming. A Japanese underwater teleoperator with a pair of slave arms, and two colour television cameras for visual transmission, has been used to screw, drill, and arc-weld. The effect of a load on the slave arms is accurately conveyed to the master units on the surface, allowing for very exact manipulation. The use of semi-intelligent robots which can perform certain tasks autonomously may provide a further solution to underwater operations. At the

moment the Marine Technology Department in Japan is working on an underwater walker which will have greater mobility than marine robots in use now.

Robots will continue to improve in terms of both intelligence and manipulative skills. A stage is likely to be reached, however, when the paradox which Norbert Wiener discussed will come into play. The paradox is embodied in the two qualities demanded of a slave: intelligence and subservience – two qualities which are not necessarily compatible. 'The more intelligent the slave is, the more he will insist on his own way of doing things in opposition to the way of doing things imposed on him by his owner. . . . The result is that in the employment of such a machine we are bound to find sooner or later that the purpose of the machine does not conform to the detailed mode of action which we have chosen for it.'[6] The problem will be to instruct the machine in such a way that there will be no discrepancy between what we have told it to do and what we meant to tell it to do. The machine has no sense of metaphor and therefore will perform its tasks in its own way.

Our task will be to instruct *ourselves* before we instruct the machine. Otherwise man may find himself in the position of King Midas: when he asked that all he touched would turn to gold, he did not intend his food to become metal, however precious. Let me end by quoting Wiener again: 'There is nothing which will automatically make the automatic factory work for human good, unless we have determined this human good in advance and have so constructed the factory as to contribute to it. If our sole orders to the factory are for an increase in production, without regard to the problems of unemployment and of the redistribution of human labour, there is no self-working principle of *laissez-faire* which will make those orders redound to our benefit and even prevent them from contributing to our own destruction. The responsibilities of automation are new, profound, and difficult.'[6]

[5] I.M. Havel and O. Štěpanková. 'The Role of Problem Solving in Cognitive Robotic Systems'. *On Theory and Practice of Robots and Manipulators*, ibid
[6] Norbert Wiener. 'The Brain and the Machine'. Sidney Hook, ed. *Dimensions of Mind*, New York University Press, 1960

Unimate, the oldest established and the most successful industrial robot, unloading a stamping press ▲
Unimates spot-welding in a motor-car assembly line. Six robots are used simultaneously on one car. ▶

144

A typical manipulator used in handling radioactive substances. It is a highly versatile mechanical linkage employing neither servomechanism nor external power. It is nevertheless suitable for light and precise work incorporating, as it does, an inherent force feedback. Michrotechna Works at Jablonac on Nisou, Northern Bohemia ▶

Mobot – a mobile robot by Hughes Aircraft Company, Los Angeles, constructed in 1960 – was designed as a substitute for humans in areas too dangerous for man to work in. With soft inflated pads on its hands and controllable pressure, it can hold delicate objects. Television cameras and microphones at the wrists provide the operator with information. It has flexible and versatile arms capable of moving 180° in any direction at each of three joints. Each arm is 6 ft (1.8 m.) long and can lift 25 lbs (11.3 kg.). Two television cameras are mounted on remotely controlled booms. Mobot can be radio- or computer-controlled, and carries one hundred command channels. The instructions can be stored on tape and repeated. Intended specifically for work in radioactive areas
▼

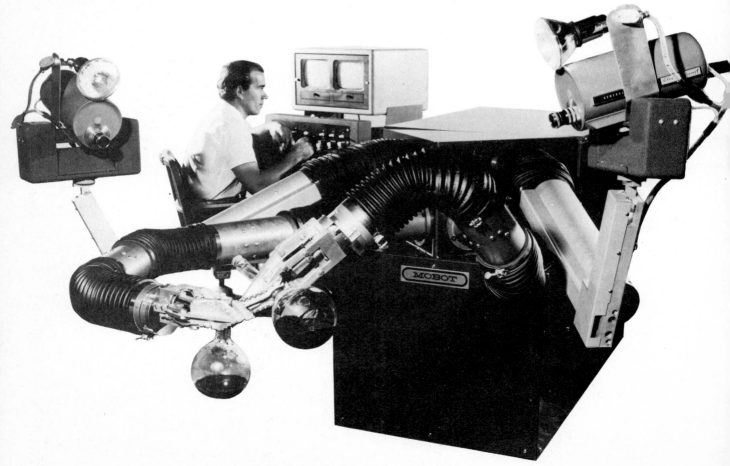

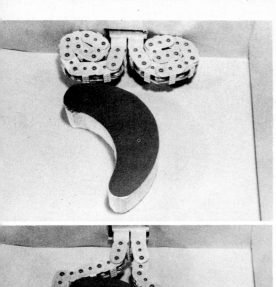

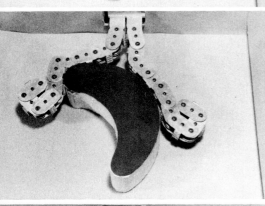

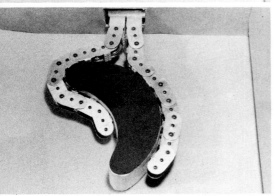

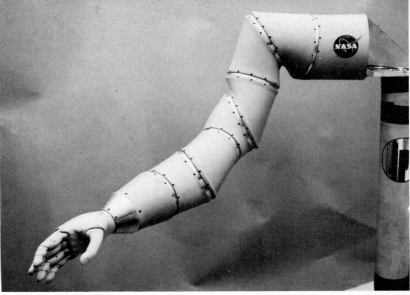

The humanoid hand attached to the JPL/Ames arm, once a part of a space suit, can handle objects of very different shapes, sizes and degrees of fragility. The thumb and all four fingers are movable and adapt themselves automatically to the shape of the grasped object. The hand can grasp with the fist or with the fingers. It represents an important stage in the development of an autonomous system with artificial intelligence capability for work on planet surfaces. For the control set-up, with master arm, manual console and computer, see colour p. 21.

▲ This soft gripper for a robot hand, inspired by the movements of the snake, is modelled on an elephant's trunk or the tentacles of an octopus rather than human fingers. Designed for the handling of fragile objects, the purpose here is to distribute the pressure uniformly about the object by coiling around it. Ideal for handling animals and human patients during medical treatment

Marks left on a clay drum show the uniform pressure exerted by the gripper. Developed at the Tokyo Institute of Technology

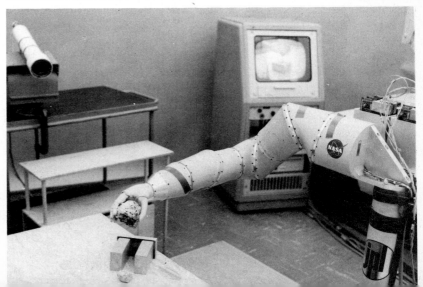

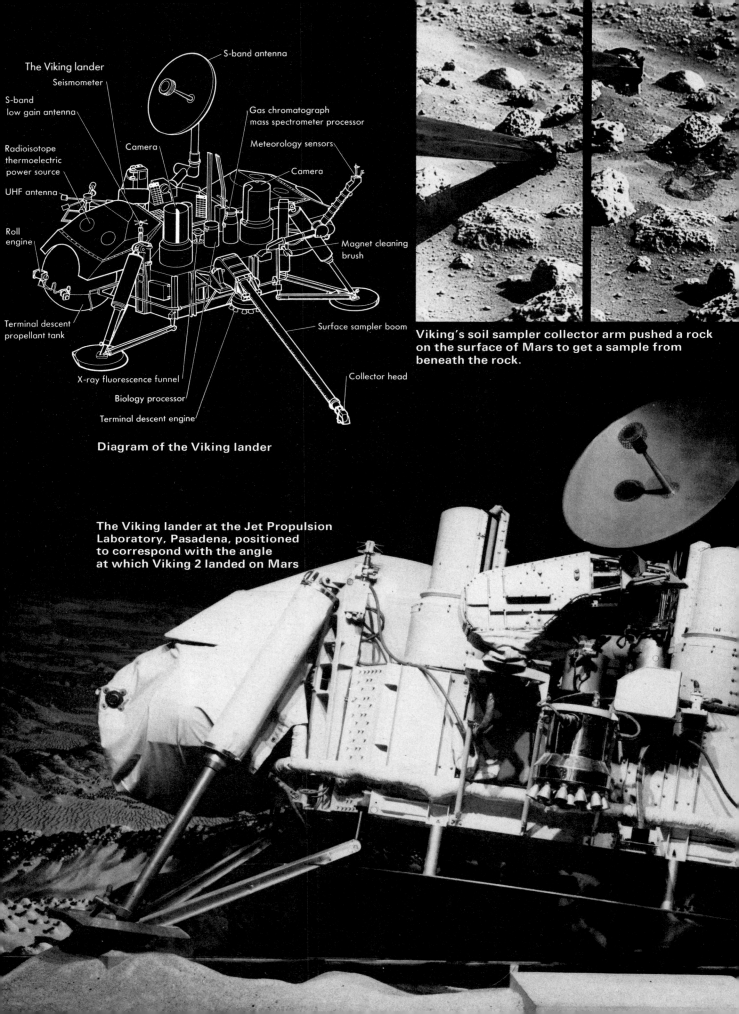

The Viking lander
Seismometer
S-band antenna
S-band low gain antenna
Gas chromatograph mass spectrometer processor
Radioisotope thermoelectric power source
Camera
Meteorology sensors
UHF antenna
Camera
Roll engine
Magnet cleaning brush
Terminal descent propellant tank
Surface sampler boom
X-ray fluorescence funnel
Biology processor
Collector head
Terminal descent engine

Diagram of the Viking lander

Viking's soil sampler collector arm pushed a rock on the surface of Mars to get a sample from beneath the rock.

The Viking lander at the Jet Propulsion Laboratory, Pasadena, positioned to correspond with the angle at which Viking 2 landed on Mars

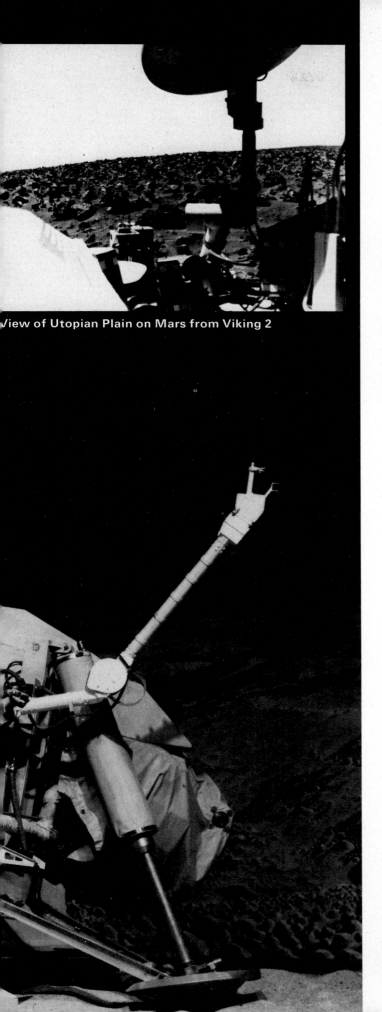

View of Utopian Plain on Mars from Viking 2

NASA Teleoperator developed at Bell Aerospace, Textron, Buffalo, NY. This model teleoperator is used in a remote control experiment to study control of space vehicles from the ground. The robot moves on an absolutely smooth deck on pin-point jets of pressurized gas, one thousandth of an inch thick. With friction thus reduced it is possible to simulate docking and other manipulations in space. Intended for installation in a spacecraft to be used for the capture and destruction of satellites, the servicing of spacecraft and astronaut rescue. It has forty digital and analog radio links with the control monitor on earth.

The first unmanned robot spaceship, Progress 1, about to dock with the space station Salyut-6/Soyuz-27, seen a few moments before link-up on 22 January 1978

Diver Equivalent Manipulator System by General Electric used for underwater operations. The arm has a reach of over 5ft (1.5 m.) and can handle weights of up to 65 lbs (29.5 kg.).

The Taylor Hitec multi-link versatile underwater manipulator can be used in conjunction with an underwater vehicle at depths of 3000 ft (914.4 m.). The arm has 7 degrees of freedom and can lift up to 200 lbs (90.7 kg.) in weight, with an overall reach of 72 in. (182.9 cm.). ▶

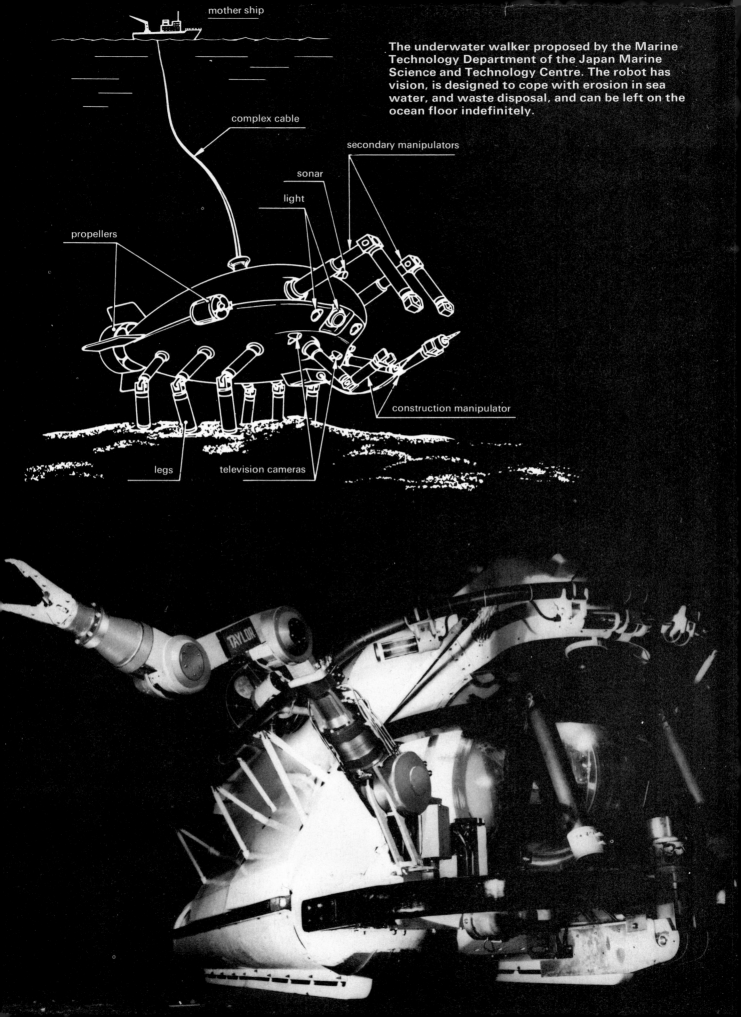

mother ship

complex cable

secondary manipulators

sonar

light

propellers

construction manipulator

legs

television cameras

The underwater walker proposed by the Marine Technology Department of the Japan Marine Science and Technology Centre. The robot has vision, is designed to cope with erosion in sea water, and waste disposal, and can be left on the ocean floor indefinitely.

Cybernetics,
the key to robot systems

'We have decided to call the entire field of control and communication theory, whether in the machine or in the animal, by the name *Cybernetics*, which we form from the Greek *kubernetes* or *steersman*. In choosing this term, we wish to recognize that the first significant paper on feedback mechanisms is an article on *governors*, which was published by Clerk Maxwell in 1868, and that governor is derived from a Latin corruption of *kubernetes*.'
Norbert Wiener. *Cybernetics*. 1948

The term was also used more than a hundred years before, in 1834, by A.M. Ampère, when he referred to the science of government as *la cybernétique*, in 'Essai sur la *Philosophie des Sciences*', Paris, 1843.

One of the most important single concepts of cybernetics, on which the functioning of every robot is based, is a servomechanism. A definition put forward by Pierre de Latil is: 'a power-amplifying mechanism intended to ensure an output bearing a functional relation to the input value.'[1] Norbert Wiener defined servomechanisms as 'systems by which we switch in an outside source of power for control purposes, such as occurs in the power steering of a truck. We call this negative feedback.'[2]

Norbert Wiener's cybernetics is partly based on observations of living creatures but without modern electronics (analog and digital computers) it would be practically impossible to make a machine that would imitate performance of a living organism.

[1] Pierre de Latil. *Thinking by Machine*. Houghton Mifflin, Boston, Mass., 1957   [2] Norbert Wiener. *I am a Mathematician*. Victor Gollancz, London, 1956

More background

Automata (discussed in 'A partial history') anticipate robots as machines that *look like* human beings or animals, and *appear to behave* like them. Cybernetic systems, aided by computers, anticipate robots as machines that may *not look like* humans but, at least partially, *do behave like* them.

Creator or source	Device	Comments
Blaise Pascal **1623–62**	Calculating machine, 1642	'I submit to the public a small machine of my own invention by means of which alone you may, without effort, perform all the operations of arithmetic, and may be relieved of the work which has often times fatigued your spirit.'
Wilhelm Gottfried von Leibniz **1646–1716**	Calculator to multiply, add, divide, and extract square roots. Built 1694	He anticipated Boole (see below) in early attempts to symbolize logical arguments in algebraic terms.

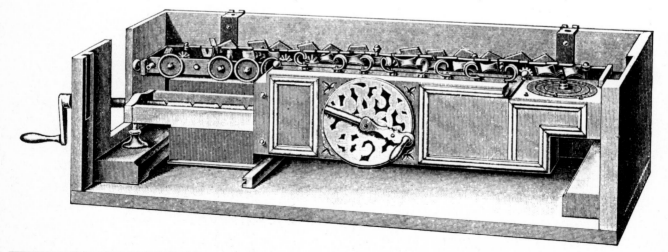

Jewna Jacobson, **clockmaker**	Calculating machine. Minsk, 1770	Used for computation of numbers up to five digits
James Watt **1736–1819**	Flyball governor, 1788	Used for regulating speed of steam engines. One of the earliest automatic control devices. (See Maxwell, below)

Luigi Galvani
1737–98

Theory of animal electricity, 1791

The first notion of the human electrical feedback system

from Galvani's *De viribus electricitatis in motu musculari*, 1792

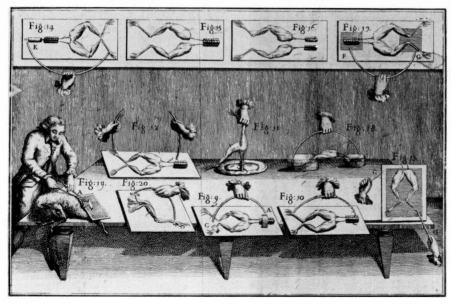

Joseph Marie Jacquard
1752–1834

Automatic loom, 1801

A train of punched cards controls the lifting of thread and thereby the fabric pattern.

Charles Babbage
1792–1871

Difference engine, 1823

Analytical engine, 1833–71

'We may say most aptly that the Analytical Engine weaves algebraic patterns just as the Jacquard loom weaves flowers and leaves.'
Lady Lovelace

**Babbage's Difference Engine was
the first of such machines
to produce tables for navigation,
insurance and astronomy
by accumulating differences**

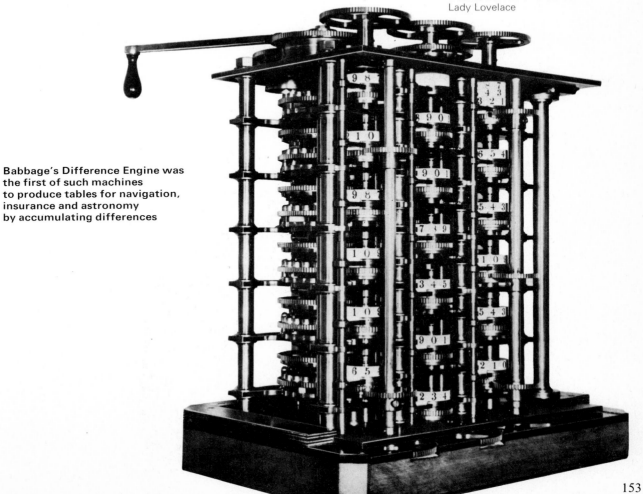

George Boole 1815–64	Boolean algebra: abstract system of postulates and symbols applicable to logical problems, universally adopted to computer use, 1854	Discussed in his 'An Investigation of the Laws of Thought', Boolean algebra leads to the binary system.
James Clerk Maxwell 1831–79	The first systematic study of feedback	His paper 'On Governors' was published in the Proceedings of the Royal Society, 1868. In his analysis of Watt's governor, he was perhaps the first to realise the significance of feedback.
Leonardo Torres Quevedo 1852–1936	Automatic electromagnetic machine capable of playing an end game of chess, 1912	At the Cybernetic Congress in Paris, 1951, G. Torres Quevedo, the son of Leonardo, is challenged by Norbert Wiener. Pierre de Latil reports that on this occasion the machine won every game.

This chess-playing automaton was probably the first decision-making machine ever to be constructed. In an interview in *Scientific American*, Leonardo Torres Quevedo said: 'The ancient automatons . . . imitate the appearance and movements of living beings, but this has not much practical interest, and what is wanted is a class of apparatus which leaves out the mere visible gestures of man and attempts to accomplish the results which a living person obtains, thus replacing a man by a machine.'

Herman Hollerith, statistician with US census office 1860–1929	Electromechanical punch-card system, 1886	Tabulated the 1890 census of a sixty-five million population in $2\frac{1}{2}$ years.
Vannevar Bush 1890–1974	Differential analyser, 1930	The first analog computer for solving differential equations
A. M. Turing 1912–54	Turing 'machine', 1936	A conceptual (never built) device investigating what kind of problems in mathematics and logic the machine could solve and what kind it could not
Thomas Ross 1916–	Machine imitating a living creature	'The first attempt to make a machine that would imitate a living creature in performance, as distinguished from appearance seems to have been suggested by the familiar test of animal intelligence in finding the way out of a maze. Thomas Ross in 1938 made a machine in America which successfully imitated this experiment. By trial and error it could 'learn' to find its way to a correct goal on a system of toy-train tracks.' W. Grey Walter, *The Living Brain*, 1938

◄ **Richard Eier's version of an artificial mouse finding its way out of a labyrinth**

Claude E. Shannon **1916–**	Demonstration how logical operations correspond to the algebra of switching (ON/OFF) circuits, 1938	'A symbolic analysis of relay and switching circuits'. *Transactions of the American Institute of Electrical Engineers*, 1938
Servomechanisms Laboratory, M.I.T.	Mathematical basis of feedback control, for high-speed aiming of artillery using radar tracking, 1939–45	
Howard Aiken **1900–73**	Automatic Sequence-Controlled Calculator, Mark I, for IBM, 1937–44	First fully automatic digital computer using punched paper tape. Could add two numbers in $\frac{1}{3}$ second.
John Vincent Atanasoff **1903–**	Automatic electronic digital computer 1940–2	Mauchly, with whom Atanasoff shared his research, patented it first and subsequently built ENIAC in collaboration with Eckert.
J. Presper Eckert, and John W. Mauchly	ENIAC – Electronic Numerical Integrator and Computer, 1943–6	First successful completely electronic computer at University of Pennsylvania with programs on switchboard circuits. Could add 5000 numbers per second.
John von Neumann **1903–57**	EDVAC – Electronic Discrete Variable Automatic Computer, 1947	Logical design of a computer capable of using a flexible stored program: program that could be changed at will without revising the computer's circuits
Josiah Macy Foundation	The first conference on cybernetics, 1947	
W. Grey Walter **1910–77**	Electronic tortoises: Elmer (electromechanical robot) and Elsie (electro-light-sensitive-internal-external), 1948	Two examples of Machina Speculatrix. The tortoises use phototropism and mechanical contact to find their way round obstacles. They feed on electricity and when their batteries require recharging, their appetite for electricity mounts, and attracted by the light placed in the 'hutch' they find their way to it, where contacts on the side of their shell engage with others in the hutch, thus closing the battery-charging circuit.

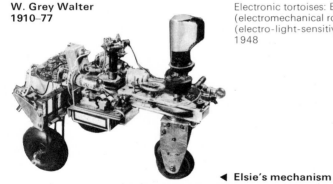

◀ Elsie's mechanism

A record of Elsie's movements during two minutes. A lighted candle on top of her provides the ▼ **luminous trail.**

W. Ross Ashby **1903–**	A homeostat, 1948	Machine with its own equilibrium which always pursues the same goal
Norbert Wiener **1894–1964**	*Cybernetics, or Control and Communication in the Animal and the Machine*, 1948	'. . . society can only be understood through a study of the messages and the communication facilities which belong to it; and in the future development of these messages and communication facilities, messages between man and machines, between machines and man, and between machine and machine, are destined to play an ever-increasing part.' Norbert Wiener. *The Human Use of Human Beings*, 1950
Bell Telephone Laboratories	Transistor, 1948	A new electronic component based on the crystal detector of early radios, which eventually replaced the valve in computer circuitry. Minute in size, it operates on a fraction of the electricity needed by a vacuum tube.
Mathematical Laboratory, University of Cambridge	EDSAC – Electronic Delayed Storage Automatic Computer, 1947–9	First practicable stored program computer with symbols to represent instructions
Servomechanisms Laboratory, M.I.T.	Whirlwind, 1947–53	First magnetic core memory – storage method allowing immediate access
	Machine tools controlled by numbers punched on paper tape, 1950s	
National Cash Register	NCR 304, 1954–7	First transistorized computer
	The term Artificial Intelligence used for the first time, 1956	At Dartmouth College, New Hampshire
	First mini-computer, 1960	
	First industrial robots controlled by electronic computers, 1960	
US Air Force	First conference on bionics, 1960	BIOlogical electroNICS. The word was coined by Dr Hans L. Oestreicher of Wright-Patterson Air Force Base, Ohio. Creating novel technological devices according to 'design principles' observed in biological organisms. Also 'biocybernetics'
Stanford Research Institute	A complete robot system	Integrated circuit computers established, 1967 Shakey, first generation, 1968

A complete robot system

Antenna

Rangefinder

Rangefinder

Television camera

Logic unit

Bump detector

Drive wheel

Drive motor

Castor wheel

The Stanford Research Institute robot vehicle Shakey (first version) was equipped with television camera, optical rangefinder and bump detectors. Powered by electric motors, its connection to a computer by cable has been replaced by radio link. Shakey can perform such tasks as locating a specific spot in its seven-room environment, avoiding obstacles, finding designated boxes and pushing them together into groups according to instructions.

University of Illinois

ILLIAC IV, 1967–71

The first parallel computer made up of sixty-four independent processing units operating simultaneously, capable of executing 200 million instructions per second

LSI chip – Large Scale Integration chip, 1968

Complete processing networks comprising thousands of transistor elements assembled on one chip

First microcomputers in general use, 1971

Complete control processors on a single silicon chip

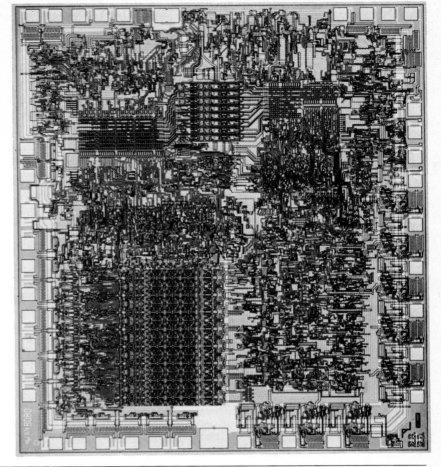

The single-chip microcomputer is a complete general-purpose digital processing and control system in one large-scale integrated circuit, made by Intel Corporation.
It measures 5.6 × 6.6 mm.

Unimation, Inc.	Robotics, as a sole business, 1972	The first firm anywhere to devote itself to the manufacture of robots
Bell Telephone Laboratories	Magnetic-bubble memory, 1974	A 250,000-bit experimental memory has been produced, opening the way for storing contents of entire encyclopedias on a single chip.
	CCD – charge couple device, 1977	Packets of electrical charge in motion, able to store more than 65,000 bits per chip
Department of Physics and Astronomy, UCL	CLIP 4 – Cellular Logic Image Processor, 1978	First digital computer to handle 9216 programmable micro-processors simultaneously

Microelectronic revolution, putting large numbers of electronic elements on silicon chips, has profoundly increased the capabilities of electronic devices. **Robert N. Noyce**

The basic functional element of a modern electronic circuit is the transistor. Microelectronic technology has made it possible to employ large numbers of them in a single circuit. **James D. Meindl**

When thousands of circuit elements are integrated on one chip, the integration is said to be large-scale. Many of these circuits not only are electrical but also follow the rules of Boolean logic. **William C. Holton**

Present memories based on transistors typically store some 16,000 bits (binary digits) on a chip. Magnetic-bubble and charge-coupled devices are providing an even higher density of information storage. **David A. Hodges**

A microprocessor is a computer central processing unit on a single chip. Currently it is associated with other chips in a microcomputing system. Now emerging, however, are complete computer systems on a single chip. **Hoo-Min D. Toong**

Microelectronics enables measuring instruments not only to make measurements but also to analyse them. It has also brought closer the fully automatic control of industrial processes and machinery. **Bernard M. Oliver**

Rates of progress in microelectronics suggest that in about a decade many people will possess a notebook-size computer with the capacity of a large computer of today. **Alan C. Kay**

Scientific American, September 1977

Artificial intelligence

Artificial intelligence[1] means artificial behaviour, or simulation of human behaviour, by computer programs and robots. Marvin L. Minsky,[2] to whom the coinage of the term is most often attributed, defines it as follows: 'Artificial intelligence is the science of making machines do things that would require intelligence if done by men.' The sort of activities involved would be: exploration, learning, pattern recognition, language translation, game playing and problem solving. Minsky adds: 'What makes for intelligent behaviour is the collection of methods and techniques that select what is to be tried next, that size up the situation and choose a plausible (if not always good) move and use information gathered in previous attempts to steer subsequent analysis in better directions.'[3]

Such behaviour is embodied in so-called heuristic programs – that is, those computer programs which use short cuts and rule of thumb procedures, as opposed to algorithmic programs, which are exhaustive as they consider every possibility available and are guaranteed to work. A heuristic procedure could be compared to that of someone setting out to solve a problem without being sure that he will succeed. A computer following such a procedure can demonstrate qualities which approximate to insight and intuition, although the computer must formalize and rationalize all data. This is unlike the way human intelligence functions but is the only way in which computers can process information.

At present, artificial intelligence programs accomplish tasks differently from a human being, sometimes better (playing nim or tic-tac-toe and winning every time), and often quicker. Since our only information-processing devices are digital computers, for which all knowledge must be expressed in terms of logical relations, the computer model is still very distant from the working brain. The work in artificial intelligence is based on assumptions that on a basic level, men and machines use the same processes for dealing with information. Real insight, however, one of the greatest human attributes, is excluded from trial and error problem solving. To analyse the problem, and split it up into components, requires an understanding of the situations in their context.

Describing a system for computer pattern recognition, Hubert L. Dreyfus lists the following necessary steps if the program is to equal human performance:[4]
1. Distinguish the essential from the inessential features of a particular instance or pattern;
2. Use cues which remain on the fringes of consciousness;
3. Take account of context;
4. Perceive the individual as typical, i.e. situate the individual with respect to a paradigm case.

In pattern recognition, computers can recognize printed and written characters with ninety-nine per cent reliability for each character, but recognizing objects placed in different positions and casting shadows is more difficult. And, as Seymour Papert has pointed out, it would be even more difficult for a computer engaged in translation to know that when 'Mary had a little lamb', she did not have it for breakfast.

These and many other problems may be resolved with the increasing complexity of the computer or with the invention of a new type of computer. Claude E. Shannon, the inventor of information theory, said: 'Efficient machines for such problems as pattern recognition, language translation, and so on, may require a different type of computer than any we have today. It is my feeling that this will be a computer whose natural operation is in terms of patterns, concepts, and vague similarities, rather than sequential operations on ten-digit numbers.'[5]

Whereas computer programs as examples of artificial intelligence proliferate, there are comparatively few robots working in conjunction with programs which can be called intelligent, and which adjust their behaviour to fit in with the goal to be achieved. Among the best-known ones is the hand/eye machine built in the 1960s at M.I.T., which consisted of an arm with a hand that could grasp, transport and assemble blocks of various shapes and sizes. A television camera viewing the area would send visual data to be analysed by computer. The arm would be given the task, say, of recognizing and picking up different shaped blocks and assembling them in a given sequence, e.g. 1. locate a pyramid; 2. locate a cube; 3. locate east face of a cube; 4. go to location of pyramid; 5. push pyramid towards the east face of the cube.

The Stanford Research Institute's Shakey (see pp. 17 and 156) was rebuilt in 1971 and given new tasks to perform. Not only did he have to assemble blocks and move them from one room

---

[1] Artificial intelligence is often referred to as AI for short, also as machine intelligence, brain modelling, and study of intelligence as computation.
[2] Marvin L. Minsky is director of the Artificial Intelligence Laboratory at M.I.T.
[3] Marvin L. Minsky. 'Artificial Intelligence'. *Information*. A Scientific American Book. W.H. Freeman, San Francisco and London, 1966
[4] Hubert L. Dreyfus. *What Computers Can't Do*. Harper & Row, New York, 1972
[5] Claude E. Shannon and Warren Weaver. *The Mathematical Theory of Communication*. University of Illinois Press, 1962

to another, he also had to open doors, go up a ramp, push blocks off the edge, and perform a number of co-ordinated manipulations with another robot, a Unimate arm.

WABOT–1, a recent experimental robot, is described as an information-powered machine with senses and limbs, which can analyse external and its own internal states, and make decisions on this basis. Constructed at the Department of Mechanical Engineering at Waseda University, Tokyo, it can walk on its two legs and perform tasks with its two artificial arms, such as pouring water from one glass into another. It has a sense of equilibrium, the hands are equipped with touch sensors, it is endowed with vision in the form of a television camera, as well as with hearing and voice, all controlled by a mini-computer. On receiving a verbal command to find a cylinder which is somewhere in the room, to pick it up and to transport it to a given location, WABOT-1 first of all acknowledges the command verbally, then its television camera scans the vicinity, locates the cylinder, after which the distance to be traversed is measured, and WABOT-1 sets off. It moves towards the cylinder but stops about half way to check all data and then proceeds further till it stops in front of the cylinder at a distance which will allow its hands to reach it. It picks the cylinder up and delivers it to the correct location.

Robots such as WABOT-1 and Shakey are experimental machines with a considerable degree of sophistication unequalled by any of the industrial robots, none of which is yet able to act autonomously. Experimental robots are learning machines and the computer programs which control them 'remember' to use methods that solved a given problem previously; they also correlate symbols with the data with which they had been associated before.

Even with learning programs, computer simulation of intelligent behaviour has many limitations, though sometimes we can find ourselves under the illusion that there is no difference between a computer and a human being. A.M. Turing, whose work on the universality of machines is one of the greatest contributions to the computer field, published in 1950 a paper entitled 'Computing Machinery and Intelligence'. He bypassed the problem of actually defining the words 'Machine' and 'Thinking' and instead examined the question of a game based on interrogation, in which the interrogator can communicate with a human being and a machine via a teletypewriter. He does not know which is which and has to decide on the basis of the answers given. The following sample of a dialogue between the interrogator and the

WABOT-1, a human-type hydraulic walking machine with bilateral artificial arms, a system for processing visual data, and a speech input–output system

The hands of WABOT-1 pouring water from one cup into another

The arrangement of sensors on the hand of WABOT-1

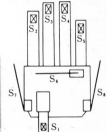

computer demonstrates that it is impossible to tell the difference:

Q: Please write me a sonnet on the subject of the Forth Bridge.
A: Count me out on this one. I never could write poetry.
Q: Add 34957 to 70764.
A: (After a pause of about 30 seconds) 105721.
Q: Do you play chess?
A: Yes.
Q: I have K at my K1, and no other pieces. You have only K at K6 and R at R1. It is your move. What do you play?
A: (After a pause of 15 seconds) R–R8 mate.[6]

There is no agreement about the future capabilities of machines. It is often claimed that computers will never be able to show true originality, or intelligence. 'At present machines can hold their own with world draughts champions and serious amateur chess players. People with this capacity would be called intelligent; but would people with *only* this capacity be intelligent?'[7] Most predictions are contradictory, but it is worth noting that already in 1950 Turing said: 'At the end of the century the use of words and general educated opinion will have altered so much that one will be able to speak of machines thinking without expecting to be contradicted.'[8]

[6] A.M. Turing. 'Computing Machinery and Intelligence'. Alan Ross Anderson, ed. *Minds and Machines.* Prentice-Hall, Englewood Cliffs, New Jersey, 1964
[7] Richard L. Gregory. 'Artificial Intelligence'. *The Fontana Dictionary of Modern Thought.* Fontana Books, London, 1977
[8] A.M. Turing. Ibid

The feelings of robots

Organisms as machines

The idea of animals as God-made automata was put forward by Descartes in 1637. He reflected that if there were a machine that looked exactly like, say, a monkey, it would not be possible to tell the difference between the imitation monkey and the real one. If, however, there were a machine which in all respects was identical to the human body, we would still have two ways of recognizing the imitation man from the real man. The first is that the machine would not be able to make use of language, verbal or gestural, to declare its thoughts.[1] The second is that although it might be able to do many things as well as man, it could not do them all, because it would lack sufficient complexity of components, or ways in which they could combine, to enable it to act 'upon every occasion of life in the same way as our reason makes us act.'[2]

A hundred years later, Julien Offray de La Mettrie extended the Cartesian animal-automaton theory so that it would apply to man. That mental life was dependent on bodily processes was already well established by the iatromechanical (from Greek *iatros:* physician) system of medicine founded on the principles of mechanics of the seventeenth century. In his *L'Homme machine* of 1747, La Mettrie discussed the problem of mind as a problem of physics.[3] He put forward many important arguments: firstly, that man is a machine composed in such a way that it is impossible to have a clear idea of it and consequently impossible to define or describe it exactly; secondly, that man is an assemblage of springs of which the brain is the principal one, the body being like a clock mechanism, ruled by time and aware of night and day; thirdly, that the machine is not immortal; and fourthly, that thought is a property of matter and that everything is made of the same material but put together differently. Two hundred years later, La Mettrie's comments have become very relevant to discussions about whether or not robots could ever be conscious.

Can machines be conscious?

Samuel Butler posed the question of whether a non-conscious machine could be improved to the point of becoming conscious. His answer was that it could, namely by speeding up its evolution, and he anticipated a future when machine consciousness would be recognized and accepted.[4] Two of the possible ways in which this can be achieved are: firstly, by adjusting the meaning of the terms 'machine' and 'conscious' so that they are encompassed within the framework of artificial intelligence; secondly, by following La Mettrie's proposal that consciousness is just a property of matter at a certain level of organization, a machine can be expected to become conscious when it becomes sufficiently complex.[5]

Discussions about the possibility of machine consciousness during the past thirty years have provoked enormously divergent responses from philosophers. Satosi Watanabe, for instance, states that consciousness is the product of the combination of certain chemicals.[6] A machine made out of protein could have consciousness but one made of vacuum tubes, diodes and transistors could not be expected to have consciousness. However, an organic artificial machine made of protein would in fact be a man-made animal and not a machine at all.

Michael Scriven comes to the conclusion that machines will never be conscious because we have come to see that a replica of a man, sufficiently exact to be conscious, is too exact still to be classified as a machine.[7]

Consciousness, however, is not a well-defined property. Hilary Putnam suggests[8] that whether robots are conscious or not is more a question of decision than of discovery. 'It seems preferable to me', he writes, 'to extend our concepts so that robots *are* conscious – for discrimination based on the "softness" or "hardness" of the body parts of a synthetic "organism" seems as silly as discriminatory treatment of humans on the basis of skin colour.' He also argues that a robot of a sufficient degree of complexity would necessarily be conscious. In his view, the discussion about consciousness centres on the assumption that robots are deterministic systems designed by someone, and that we are not. But, should we discover one day that we too are artifacts, and that everything we do and say has been anticipated by our super-intelligent creators, it could follow that we had no consciousness either.

Programming in man and machine

'Once upon a time man set out to construct a human brain. He had at his disposal a number of marvellous machines and he had invented several others: calculating machines dealing with 186 types of problem, machines capable of the operations of logical thought, those that could predict the future on the strength of information received from sensory mechanisms, machines of universal memory which could deal with the sum of human knowledge, a machine capable of playing chess or bridge or doing crossword puzzles. Thus he had 83,287 infallible machines.

Drawing by Masami Miyamoto

The following definition of a man
was written by R. Buckminster Fuller in 1938.
'A self-balancing, 28-jointed adapter-base biped;
an electrochemical reduction plant, integral with
segregated stowages of special energy extract in
storage batteries for subsequent actuation of
thousands of hydraulic and pneumatic pumps
with motors attached; 62,000 miles of
capillaries . . . The whole, extraordinary complex
mechanism guided with exquisite precision from
a turret in which are located telescopic and
microscopic self-registering and recording range
finders, a spectroscope, etc.; the turret control
being closely allied with an air-conditioning
intake-and-exhaust, and a main fuel intake . . .'

When he had set them up in an immense building,
he proudly pointed to the building where
everything could be accomplished without human
aid and claimed to have made a brain that could
do all that it is possible for the human intellect to
accomplish. Everyone wished to test the machine.
The wise men asked it questions which they alone
knew how to answer and the machine gave the
correct response. The masters of chess or of
bridge challenged the machine and it beat them.
But one day a small boy turned up saying that he
wanted to play "snakes and ladders" with the
machine. "Snakes and ladders?", cried the inventor, "but I never thought of a machine for
playing that." '

The moral of this story by Pierre de Latil[9] is that
the machine has only the structure designed for it
by man, and that the range of its activities is
strictly limited, even though there may be a great
many of them. Thus it can do only those things it
has been programmed to do. In this way it is like a
human being. If the story of Genesis is true,
suggests J.J.C. Smart,[10] then Adam and Eve, as
artifacts made by God, were robots which were
given their programs in the form of genes, and
subsequently everything that man has done
demonstrates this programming. The difference
between a robot programmed by man and man
programmed by God, according to Ninian
Smart,[11] is that a robot can be given a number of
programs which one can change, but man has
been condemned to one set of programs for ever.

Noam Chomsky talks of man being pre-
programmed for the accomplishments which he is
able to attain. He suggests that for the acquisition
of language, for instance, there is no other
explanation. How, otherwise, could human
beings master something as difficult as a language
so quickly at a very early age. He puts forward the
theory that all human languages must have a
basic structure in common and that it corresponds to this pre-programming. The genetic
program which establishes a set of constraints 'is
what provides the basis of our freedom and
creativity. If we were plastic organisms without
extensive pre-programming, the state that our
mind achieves would, in fact, be a reflection of the
environment, which means that it would be
extraordinarily impoverished.'[12] But whereas
pre-programming may be the source of creativity,
at the same time it undoubtedly limits the leaps of
imagination required for solving all sorts of
intractable problems.

[1] Though he could imagine a machine that would react verbally to mechanical changes – 'qu'elle crie qu'on lui fait mal, et choses semblables'.
[2] René Descartes. *Discourse on the Method of Rightly Conducting the Reason, and of Seeking Truth in the Sciences*, part 5, 1637
[3] Julien Offray de La Mettrie. *L'Homme machine*. Elie Luzac, Leyden, 1747
[4] Samuel Butler. *Erewhon, or, Over the Range*, London and Edinburgh, 1872
[5] La Mettrie. Ibid
[6] Satosi Watanabe. 'Comments on Key Issues'. Sidney Hook, ed. *Dimensions of Mind*. New York University Press, 1960
[7] Michael Scriven. 'The Mechanical Concept of Mind'. *Mind*, vol. lxii, no. 246, 1953
[8] Hilary Putnam. 'Robots: Machines or Artificially Created Life?' *Mind, Language and Reality*. Philosophical Papers, vol. 2. Cambridge University Press, 1975
[9] Pierre de Latil. *Thinking by Machine*. Houghton Mifflin, Boston, Mass., 1957
[10] J.J.C. Smart. 'Professor Ziff on robots'. *Analysis*, vol. xix, no. 5, 1959
[11] Ninian Smart. 'Robots Incorporated'. *Analysis*, vol. xix, no. 5, 1959
[12] 'Noam Chomsky on the Genetic Gift of Tongues'. Edited version of a discussion between Noam Chomsky and Bryan Magee, in the BBC2 television series 'Men of Ideas: Creators of Modern Philosophy'. *The Listener*, London, 6 April 1978

'The Feelings of Robots'

Under this title, Paul Ziff wrote in 1959[13] a paper in which he examined some of our attitudes to robots, concluding that in fact robots could not have feelings. Robots as computing machines are capable of performing countless laborious, repetitive, or tiresome tasks and we expect them to have the attributes of machines, even when they are made to look like men with appropriate limbs, clothes, and masks. The reason why a robot cannot have feelings is that we do not yet know how to isolate feelings and program them into a piece of hardware. When a robot looks tired we may safely say that it has been constructed to look, under the given circumstances, like a man who is tired. Since they are just machines, they can be broken but not dead, and malfunctioning but not tired. A robot can kill people but not murder them, it can paint a picture but not create a work of art. Ziff argues that feelings are something to be programmed rather than the result of the combination of programming and circumstances.

However, Michael Scriven puts forward a very different view. Already in 1953 he had said: 'I now believe that it is possible to construct a supercomputer so as to make it wholly unreasonable to deny that it had feelings.'[14] To put this to the test would be difficult since it may be impossible to decide whether the robot is feeling pleased when it looks pleased, or whether it has been programmed, as Ziff suggests, to look like a man who is pleased under certain conditions. Scriven's solution is to program the robot so that it is impossible for it to lie. If it already has the same performing abilities as man and, further, has been fed with all the works of great poets, novelists, philosophers and psychologists so as to understand what we mean by 'feelings', it could be asked if it has them. If it answered 'yes' we would have no more reason to doubt its answer than that of anyone else. If we wanted to know whether an artifact could be a person, then the first robot to answer 'yes' would qualify.[15]

Civil rights for a robot-person

If a robot meets the criteria of consciousness, which means having feelings, thoughts, attitudes and personality, and can be referred to as a robot-person, then what sort of civil rights must it be given?

William G. Lycan, delivering in 1972 a lecture on the civil rights of robots,[16] used the term robot-person, implying that since our definitions of the differences between people and robots are awkward, it could be useful in principle to treat the robot of the future as a person. The robot-person is an android endowed with any known mental state desired. It is manufactured but otherwise indistinguishable from man. He chose as his example a robot called Harry who 'can converse intelligently on all sorts of subjects, play golf, write passable poetry, control his occasional nervousness pretty well, make passionate love, prove mathematical theorems, attend political rallies with enthusiasm, show envy when outdone, throw gin bottles at annoying children, etc.'

'We are not logically compelled to call Harry a person; at best it would be natural to call him one', once it has been agreed that, as far as one can establish, Harry is conscious and capable of feeling pain, pleasure, and so on. There are two differences between Harry and a human being: Harry is an artifact and he is made partly or entirely of hardware, but as Lycan says: 'if we object to racial and/or ethnic discrimination in our present society, we should object to discrimination against Harry on the basis of his birthplace.' Intermediate cases of robots, before they become fully-fledged persons, cannot morally be neglected either. Since we cannot be sure at which point sentience develops into emotion, it is best not to decide to treat them differently. They could have the same rights as children or animals.

Problems are bound to develop later. Lycan ends his lecture with remarks on a category of superior machines which are *super-beings.* It is conceivable that 'two or more extremely intelligent and sensitive beings who were created by humans could themselves build a super-being who was so superior in every way to humans that it could never have been created by humans themselves.' Such beings would presumably have more rights than humans, but it is difficult to imagine what they might be.

[13] Paul Ziff. 'The Feelings of Robots'. *Analysis,* vol. xix, no. 3, 1959
[14] Michael Scriven. Ibid
[15] Michael Scriven. 'The Compleat Robot: A Prolegomena to Androidology'. Sidney Hook, ed. *Dimensions of Mind.* New York University Press, 1960
[16] William G. Lycan (Associate Professor, Department of Philosophy, Ohio State University). 'The Civil Rights of Robots'. Lecture at Kansas State University, October 1972

Drawing by Hans Küchler

Robots, ourselves, and the future

A bored society trying to combat pollution of information, having token jobs or no jobs, and watching the new machine-species, which had been passive for centuries, become active and proliferate – such is one of the many possible descriptions of the future when automation reaches a certain stage of development. Already in 1962, Oliver G. Selfridge had proposed a way of alleviating some of the social consequences of this vision of the future with his suggestion (to be taken *cum grano salis*, no doubt) that 'being unemployed is a harder job than most and should be rewarded accordingly. If the pay were high enough, people would willingly give up half their pay in order to have a job.'[1]

The two assumptions that have been made so far are that people have an innate biologico-ethical need to work, and that the machines are lying in wait for the opportunity to rise up, grow, multiply and take over more and more of the world, both physically and psychologically.

Visions of the future have little bearing on reality when it finally arrives and it is possible that man will cease to associate work with necessity, and that machines will turn out to be something quite different from what we have imagined. Edward de Bono once suggested that robots of the future might simply consist of bags of protoplasm capable of assuming any shape, which will perform whatever tasks may be required. The main difference between his and most ideas about robots is that they will be soft rather than hard.

The fact that historical events do not actually take place in the way we anticipate them may be the reason that some of the otherwise plausible predictions which have been made have not materialized. Had our priorities not changed during the 1970s, with the new awareness of the necessity to conserve energy resources, and had the funds for research in artificial intelligence and robotics been increased, as had been expected, Marvin Minsky's prediction for 1978 might have come closer to realisation than it did. In 1970 he said: 'In three to eight years we will have a machine with the general intelligence of an average human being. I mean a machine that will be able to read Shakespeare, grease a car, play office politics, tell a joke, have a fight. In a few months it will be at genius level and a few months after that its powers will be incalculable.'[2]

In 1964, Professor M.W. Thring made a prediction about the domestic robot. In his article 'A Robot About the House',[3] he wrote about life in 1984, a time by which he anticipated man would work much shorter hours and 'the great majority of housewives will wish to be relieved completely from the routine operations of the home such as scrubbing floors, or bath or cooker, or washing clothes and washing up, dusting or sweeping, or making beds'. When talking to some women about domestic robots, he reports that the immediate reaction of ninety per cent of them was to wonder how soon they would be able to buy such a robot. By 1984, if not before, the tasks mentioned by Professor Thring will no longer be done by people, but not because they will be done by robots: floors will be scrubbed by scrubbing machines, washing up and washing will continue to be done by machines, beds will not be made because bedding will have changed so that beds will not need making, and machines absorbing dirt from the atmosphere will make dusting redundant. Robots will have little involvement with domestic work because it is unlikely that at a time of recession investment will be made to develop the sort of machinery which could do more or less what a person can do about the house. Thus Professor Thring has changed his ideas about the development of robotics and he considers now that telechirics (from Greek – hands at a distance), already used in nuclear work, space research and on the ocean floor, may be used for mining coal and cultivating the desert, and be of greater value to humanity than some sophisticated robots dusting bookshelves.

Whereas robots in the home are not likely to be given priority, they will continue, in increasing numbers, to work in industry. In their wake they will bring more new developments in industrial architecture designed to accommodate robot operations and automated delivery services working under the minimum of human supervision. As industrial robots cannot function in a disordered environment, special arrangements will have to be made for them not only in new industrial premises but also in public buildings where they might be used for surveillance, rubbish disposal and for dispensing food and drink.

In medicine, robots are likely to be used as receptionists and interviewers: it has already been noted in the medical press that patients give a more candid account of their symptoms and medical history when talking to machines. In schools, robots will be used as educational aids

---

[1] Oliver G. Selfridge. 'What to do about Automation'. I.J. Good, general ed. *The Scientist Speculates: an Anthology of Partly-baked Ideas*. Heinemann, London, 1962
[2] Marvin Minsky quoted in: Carole Spearin McCauley. *Computers and Creativity*. Praeger, New York, 1974
[3] M.W. Thring. 'A Robot About the House'. Nigel Calder, ed. *The World in 1984*, vol. 2. *New Scientist*, London, 1964; and Penguin Books, London, 1965

and to entertain children: their presence would accustom children to working with intelligent machines in the future. According to some predictions, there may also be a very specific area where, for lack of human volunteers, machines will be used to keep human beings company, to listen encouragingly to what they have to say, and to boost their confidence with compliments.

The great number of amateurs making robots for fun guarantees that complicated toys, already called pet-robots, will come on the market. The US Robotics Society, with a large membership of robot makers, has put forward details of such a robot which is well within the range of our present capabilities:[4] it will weigh between 25 and 40 lbs (11.3–18.1 kg.), and move at the speed of $\frac{1}{2}$ to 2 mph (0.8–3.2 kph); it will be roughly spherical, have large brown eyes and a high forehead; it will look, smell and taste like a friendly mammal, and feel warm; it will purr, click and chortle in a random fashion, follow people around, respond with pleasure to the sound of a dinner bell and show delight at being 'fed', i.e. plugged in for a battery recharging. Apart from all this, it will guard the house against 'amateur' burglars. 'Imagine that you are a burglar. You have entered an apparently empty house and are rummaging around for the cameras when something you cannot quite identify hastens into the room to announce in a slightly hysterical voice: "I think it's only fair to warn you that there are rattlesnakes loose in here." Then the thing groans miserably and loudly and settles down to follow your moves with its sensors. From time to time, a rattlesnake rattle is heard in the room, accompanied by little gasps. This might inhibit you from reaching into dark corners and unless you are a very hardened thief – you will go away.' (Robert Rossum does not take into consideration the possibility that someone may wish to steal not his cameras but his rattlesnake!)

We are able to say something more definite about the future use of robots in industry, in schools, in medicine, and in the nursery, because those sorts of robots are already being used. On the other hand, the prediction that robots will run the Stock Exchange or Sotheby's is as likely to come true as not. There are no theoretical reasons at the moment why this should not happen, but in a hundred years' time it may become obvious why it did not.

The sort of future predictions we cannot possibly quarrel with are those we will have no means of verifying. Many of the most interesting of them are about a 'perfect' machine. Usually the date when such a machine would be built is not given, although when I.J. Good first described the Ultra-Intelligent Machine in 1962, he predicted it for 1978.[5] Its most important feature was that it had 'something of a biological structure resembling that of the brain'. It was predicted that it would develop moderately slowly in machine terms, starting with a brain resembling that of a small lizard which in due course would develop into one with the intelligence of a human baby. With a comparatively small increase in cost, one could reach the level of the baby Newton. It could then be educated, and in due course it could design another, more powerful, machine. Such machines would eventually look after the problems of science and government, as well as those of overpopulation and unemployment, of which they would unwittingly have been the cause, having eliminated both disease and work. I.J. Good also considered the possibility that a machine, being selfish, might not wish to replace itself by designing a better model. This, he proposed, could be overcome by allowing the machine to continue to improve itself. In due course, the machine would take over world government, and Good guessed that this would all come to pass in 1978.

In 1972, Inspector Watson of the Special Branch told Stefan Themerson,[6] in great confidence, that Dr Good's Ultra-Intelligent Machine was already being built. The Ultra-Intelligent Machine, or UIM for short, was to be the last invention of mankind since 'she' would be able to do everything better than man. The debate between the two men centres on the fact that the UIM, if she is to arrive at an understanding of the world, may need to acquire some concepts which are unknown to man, and the question is where she is to find them.

It is understood that, in due course, the UIM would inevitably become another God, or Goddess, built in our own image. The reason for being concerned about her programming, however, is that it is vital to build a God men can understand. Furthermore, it is even more important to build one that could not, under any circumstances, become vicious. The main question Inspector Watson asks is: 'What sort of a gadget ought to be built into her hardware? . . . what sort of programming ought to be put into her software – to give her, so to speak, a moral nature?' The Inspector receives three pieces of advice. First, 'As she is a *logical* machine, it's obvious that you can't feed any ought-arguments into her. Because there is no logical argument to tell her why one ought not to kill or cheat or oppress or tyrannize.' Second,' . . . the machine is going to be rational. Therefore: Don't put any beliefs into her software. And especially the beliefs in the happiness of the greatest number.'

The pursuit of the greatest happiness for the greatest number of some particular group of people could lead to the belief that everyone else could be disposed of. Third, she should not read Plato first – meaning that she should not be allowed to develop a syndrome of classifying things before she has had time to build her own vision of the world. ' . . . if you want your Ultra-Intelligent Machine to be more intelligent than we are, don't let her inherit our classificatory madness. Let her find her own way of putting things together. Let her make her own conceptual revolution. Let her make a proper job of it. Give her a high-speed sub-assembly fitted with a battery of 100,000 calf-brains, or 100 million cockroach brains, but let her consider each and every characteristic as equal and as important as another.'

Furthermore, she should be programmed not with love but with knowledge of how things are: 'Don't put Love into her software; put Biological Altruism into her hardware' is the final advice Stefan Themerson gives to Inspector Watson. The moral nature of perfect machines is usually taken for granted, and this is the only example of a machine destined to rule the world, or the universe, or the cosmos, to be educated so that she becomes fit to perform such a task.

The role of the machine as God is ultimately to preserve humanity by methods which are acceptable to man. It is assumed that by taking certain precautions – the three pieces of advice of Stefan Themerson and the three laws of robotics of Isaac Asimov (see p.41) – the machine will conform to this model.

In Isaac Asimov's story[7] of a machine which becomes God, it has already begun to perform its task as a vast computer, Multivac, spread over many miles and trapping the sun's energy to provide all necessary power for the earth. Two half-drunk Multivac technicians, worried about the energy of the sun eventually running out, ask the computer if it is possible to decrease the entropy of the Universe. Multivac's reply is: 'Insufficient data for meaningful answer'. Centuries later, each planet has a vast AC (the only letters left of the original Multivac), and every spaceship is a miniature AC; once more some space travellers ask the same question and once more receive the answer 'Insufficient data for meaningful answer'. The same question will be asked and the same answer given over and over again. Millions of years later, mankind, now immortal and occupying the Milky Way, is looking for further galaxies to colonize, and although each new AC is better and bigger than the previous one, it cannot answer the question.

Hundreds of millions of years later, mankind occupies all the galaxies and no longer has physical bodies, but somehow can radiate the individuality of its members through energy. AC is a small two-foot globe which partly exists in a multi-dimensional hyperspace and is difficult to see. Even this AC cannot answer the question. Gradually mankind loses its individual identity and becomes a fusion of trillions of human beings and, as the stars are dying, man still asks the AC, which by now totally exists in hyperspace, if anything can be done about entropy. Eventually mankind fuses with AC, matter and energy come to an end, space and time no longer exist. AC now exists solely for the sake of finding an answer to the ultimate question and can only release its consciousness when all data has been collated and organized. During this timeless process of correlating all past information, AC has learned how to reverse the process of entropy, but there are no men to give the answer to. 'No matter. The answer by demonstration would take care of that too. For another timeless interval, AC thought how best to do this. With great care AC organized the program. The consciousness of AC encompassed all of what had once been a universe and brooded over what was now chaos. Step by step, it must be done. And AC said, "Let there be light!" And there was light.'

[4] Robert Rossum. 'Robots as Household Pets'. *Interface*, vol. 2, no. 5, Cerritos, California, April 1977. (Robert Rossum is the collective nom-de-plume for members of the United States Robotics Society.)
[5] Irving John Good. 'The Social Implications of Artificial Intelligence'. I.J. Good, general ed. *The Scientist Speculates: an Anthology of Partly-baked Ideas.* Heinemann, London, 1962
[6] Stefan Themerson. *Special Branch.* Gaberbocchus Press, London, 1972
[7] Isaac Asimov. 'The Last Question'. *Science Fiction Quarterly,* Holyoke, Mass., November 1956

Jean-Michel Folon. *Robot* **(detail), undated, gouache on paper, 10½×8 in. (26.7×20.3 cm.)**

Acknowledgments

(continued from p. 4)

My warmest thanks go to the following for responding with great generosity to requests for material, advice, explanation, instruction, translation, and help of every sort:

Clayton Bailey, Wonders of the World Museum, Port Costa, Calif.; Gwen Barnard, London; Monika Beisner, London; Dr A.K. Bejczy, Jet Propulsion Laboratory, Pasadena; Dr Margaret Boden, School of Social Sciences, University of Sussex; Tim Brown, University of London Institute of Education; Gordon Clough, London; Mr D.W. Collins, Hugh Steeper Ltd, Roehampton; John Collins, BOC, London; Brian Davies, Dept of Mechanical Engineering, University College, London; Prof. Ron Davies, BRADU, Roehampton; Bob de Moor, Brussels; Thomas Dick, Bio-Engineering Unit, Princess Margaret Rose Orthopaedic Hospital, Edinburgh; Don Flowerdew, Royal College of Art, London; Jean-Michel Folon, Burcy, France; Prof. Michael Freeman, Management Sciences, Bernard Baruch College, New York; Stanisław Frenkiel, University of London Institute of Education; Mr Jean Gimpel, London; Godfrey Goodwin, London; Dr Tim Gordon, MRC, London; Eric Haron, London; David L. Heiserman, Columbus, Ohio; Mr J.-J. Indermühle, Cultural Counsellor, Swiss Embassy, London; Dr Hidé Ishiguro, Dept of Philosophy, University College, London; Eric James, London; Japan Industrial Robot Association, Tokyo; Irene King, Rhyl Primary School, London; Ewa Kuryluk, London; Bruce Lacey, London; Jerome Y. Lettvin, M.I.T., Cambridge, Mass.; John and Pamela Lifton-Zoline; Yoshiko Lloyd, London; Dr Alan Mackay, Dept of Crystallography, Birkbeck College, London; Peter Marginter, Austrian Institute, London; Prof. L. Maunder, Faculty of Applied Science, Newcastle upon Tyne; Colin McGinn, Dept of Philosophy, University College, London; Prof. Pamela McCorduck, Dept of English, University of Pittsburgh, Pa.; Dorothy Morland, London; National Engineering Laboratory, Glasgow; Peter Nicholls, London; Deanna and Guy Petherbridge, London; Jerome R. Ravetz, Dept of Philosophy, University of Leeds; Larry Rosen, New York; Hilary Rubinstein, A.P. Watt, London; Prof. Claus C. Scholz, Vienna; Prof. Ali Seireg, Dept of Mechanical Engineering, Madison, Wisc.; John Sladek, London; David M. Smith, Dept of Mechanical Engineering, University College, London; Yolanda Sonnabend, London; Robin Spencer, Krazy Kat Archive, University of St Andrews, Scotland; Stationers' Company's School, London; George Steiner, Cambridge; Peter Steuer, Pfeggingen, Switzerland; Herman Swart, Amsterdam; Franciszka Themerson, London; Alvin Toffler, New York; Unimation Inc., Danbury, Conn.; Donald A. Vincent, Robot Institute of America, Dearborn, Mich.; Mr M. Vitali, Limb Fitting Centre, Roehampton; Maria Vivié, London; Dr P.A. Wallace, Division of Research in Medical Education, University of Southern California; John Willats, Faculty of Art and Design, North East London Polytechnic; Arnold Zuboff, Dept of Philosophy, University College, London.

I should also like to express my gratitude to the team at Thames and Hudson who worked on this book.

Photographs used in the illustrations are by courtesy of the following (*t* top, *m* middle, *b* bottom, *l* left, *r* right):

AMF Electrical Products Development Division, London 143 *b*. Collection the artist 57. Copyright ASEA 142 *tr, mr, b*. Associated Press 149 *br*. Universitäts-Bibliothek, Basle 123; 131 *bl, mr*; 137. Bell Aerospace Textron, Buffalo, New York 149 *m*. Bildarchiv Preussischer Kulturbesitz, Berlin 14 *tr*; 114. The Bettmann Archive Inc. 167. Bodleian Library, Oxford 11 *tr*. Collection Günter Böhmer, Munich 91. Courtesy, Museum of Fine Arts, Boston. Harvey Wetzel Fund 11 *tl*. BBC copyright photograph 24 *b*; 28 *r*. British Library, London 10. Brookhaven National Laboratory 18. Carl Byoir & Associates, Inc. 146 *b*. Camera Press, London 116 (photo by John Drysdale); 120 *t* (photo by Jonathan Harsch); 120 *bl*. Cincinnati Milacron, Ohio 142 *tl*. Becky Cohen 60. Crown Copyright 125 *t*. CTK-files 39 *t, m, br*; 146 *t* (Viktor Lomoz). J.M. Dent & Sons Ltd 102 *tl, tr*. Depero Museum, Rovereto 48. Anthony d'Offay 5; 53. Tony Evans 58. Reprinted by permission of Faber & Faber Ltd 101 *l*. Editions Georges Fall 43. Jean-Michel Folon 165. Franklin Institute, Philadelphia 15 *b*. Dr Michael J. Freeman, Baruch College of the City University of New York 108. Fujimura 22. Gaberbocchus Press, London 130. General Electric Company, Philadelphia 20; 134; 136; 139 *b*; 150. Estate of George Grosz, Princeton, New Jersey (Photo Galerie Nierendorf, West Berlin) 55. Illustration from *The Iron Man*, written by Ted Hughes, illustrated by George Adamson. Drawings copyright © 1968 by George Adamson. By permission of Harper & Row Publishers, Inc. and Faber & Faber Ltd 101 *l*. Historical Pictures Service, Inc. 113 *tr*. Hitachi Ltd, London 143 *t*. IBM United Kingdom Ltd 153 *b*. Photo Edward Ihnatowicz 61; 138. Intel Corporation (UK) Ltd, Oxford 157. Internationale Bilderagentur 81 *b*. Jet Propulsion Laboratory, California Institute of Technology 21; 147 *r*. 'Dynamics of Change' © 1977 Kaiser Aluminium & Chemical Corporation 161. Keystone 93 *b*; 120 *br*. Mrs Frederick Kiesler 38 *t*. Krazy Kat Archive, St Andrews 95; 98 *t*; 99 *t*. Kunsthistorisches Museum, Vienna 12 *t*. Jean-Philippe Lenclos, Paris 23. Louvre, Paris 132 (photo Giraudon). Manning Bros 139 *t*. Mego Corp., New York 96. Metro–Goldwyn–Mayer, Inc. 65 *t*; 69. Ian Miller 85. Musée National des Techniques, Conservatoire National des Arts et Métiers, Paris 13 *t*. NASA 148 *b, tr*; 149 *tl*. National Film Archive 29 *t*; 64; 65; 66 *m, tl, b*; 67; 68; 69. Musée d'Art et d'Histoire, Neuchâtel 14 *tl, tm, lm*. Collection the Museum of Modern Art, New York 49 (Gift of Jean and Howard Lipman); 54 (Gift of A. Conger Goodyear). Seiji Otsuji 73 *t*. Popperfoto 81 *t*. Print Mint, California 87. Quasar Industries, Inc., Rutherford, New Jersey 115 *r*. Radio Times Hulton Picture Library 13 *m*; 37; 75. Reckitt Industrial Division, High Wycombe 117. Roger-Viollet 6. The Seabury Press, Inc. 44; 45 *t*. Brian Shuel 23 *b*; 97; 98 *b*; 107. Sierra Engineering Company 111 *tl, b* (photo Transaero). René Simmen Verlag, Zurich 31 *t*; 72; 162; 166. Stanford Research Institute, California 17; 127; 156. Photo Moderna Museet, Stockholm 51. Bilderdienst Süddeutscher Verlag 63; 80; 84; 92; 121. Syndication International 115 *l*. Taylor Hitec Ltd, Chorley, Lancs. 151 *b*. Tate Gallery 59 (photo Marlborough Fine Art Ltd). Technisches Museum für Industrie und Gewerbe, Vienna 13 *bl*; 154 *b*. Tokyo Institute of Technology 147 *l*. Dept of Mechanical Engineering, Tokyo University 133 *t*. Desmond Tripp 155 *b*. By courtesy of Twentieth Century–Fox Film Company Ltd 24 *t*. By courtesy of Twentieth Century–Fox Television 28 *l*. Unimation Inc., Danbury, Connecticut 19 *b*; 145. United Artists Corporation Ltd 67 *t*. U.S. Army Photograph 119 *l*. Universal 64; 65 *b*. University College, London 123 *br*. Victoria and Albert Museum, London 15 *tr*; 93 *t*. Waseda University, Tokyo 159. Oliver Watson 131 *t*. Gerry Webb of Science Fiction Consultants (UK) 42; 47; 86. By courtesy of the Wellcome Trustees 131 *br*; 137 *br*.

Drawing by Hans Küchler

Selected bibliography

The following is a general list of books concerned with robots and allied issues. Readers interested in robots in science fiction should consult item 2 for a comprehensive reading list; a good bibliography on artificial intelligence is contained in item 6; for up to date information on robots in industry, research, and medicine, the best source is the proceedings of the latest International Symposium on Industrial Robots or International Conference on Industrial Robot Technology, organized by such bodies as the Japan Industrial Robot Association, Robot Institute of America, and the British Robot Association.

The editions of the books listed are not necessarily the earliest but the most readily available.

1. **Alan Ross Anderson,** ed. *Minds and Machines.* New Jersey, 1964
2. **Brian Ash,** ed. *The Visual Encyclopedia of Science Fiction.* London, 1977
3. **Isaac Asimov.** *I, Robot.* London, 1972
4. **Isaac Asimov.** *The Rest of the Robots.* London, 1972
5. **Robert M. Baer.** *The Digital Villain.* California, 1972
6. **Margaret A. Boden.** *Artificial Intelligence and Natural Man.* Hassocks, Sussex, 1977
7. **Samuel Butler.** *Erewhon, or, Over the Range.* London 1960
8. **Martin Caidin.** *Cyborg.* London, 1973
9. **Nigel Calder.** *Robots.* London, 1958
10. **Karel Čapek.** *R.U.R. – Rossum's Universal Robots,* in The Brothers Čapek. *R.U.R. and the Insect Play.* Oxford, 1973
11. **Michel Carrouges.** *Les Machines Célibataires.* Paris, 1976
12. **Alfred Chapuis and Edouard Gélis.** *Le Monde des automates.* Paris, 1928.
13. **Alfred Chapuis.** *Les Automates dans les œuvres d'imagination.* Neuchâtel, 1947
14. **Alfred Chapuis and Edmond Droz.** *Automata – a Historical and Technological Study.* Neuchâtel, 1958
15. **P. E. Cleator.** *The Robot Era.* London, 1955
16. **John Cohen.** *Human Robots in Myth and Science.* London, 1966
17. **Alfred J. Cote** jr. *The Search for the Robots.* New York and London, 1967
18. **Pierre de Latil.** *Thinking by Machine.* Boston, Mass., 1957
19. **Hubert L. Dreyfus.** *What Computers Can't Do – a Critique of Artificial Reason.* New York, 1972
20. **Charles and Ray Eames.** *A Computer Perspective.* Cambridge, Mass., 1973
21. **Robert Escarpit.** *The Novel Computer.* London, 1966
22. **Edward A. Feigenbaum and Julian Feldman,** eds. *Computers and Thought.* New York, 1963
23. **I. J. Good,** general ed. *The Scientist Speculates – an Anthology of Partly-baked Ideas.* London, 1962
24. **Mary Hillier.** *Automata and Mechanical Toys.* London, 1976
25. **Sidney Hook,** ed. *Dimensions of Mind.* New York, 1960
26. **Stanisław Lem.** *The Cyberiad.* London, 1975
27. **Warren S. McCulloch.** *Embodiments of Mind.* Cambridge, Mass., 1965
28. **Gustav Meyrink.** *The Golem.* London, 1928
29. **Hilary Putnam.** *Mind, Language and Reality.* Philosophical Papers, vol. 2, Cambridge, 1975
30. **Bertram Raphael.** *The Thinking Computer – Mind Inside Matter.* San Francisco, 1976
31. *Information* – a Scientific American Book. San Francisco, 1966
32. **Mary Shelley.** *Frankenstein.* London, 1976
33. **Clifford D. Simak.** *City.* New York, 1976
34. **René Simmen,** ed. *Der mechanische Mensch.* Zurich, 1967
35. **Carole Spearin McCauley.** *Computers and Creativity.* New York, 1974
36. **Rolf Strehl.** *The Robots are Among Us.* London, 1955
37. **Stefan Themerson.** *Special Branch.* London, 1972
38. **Pierre Versins.** *Encyclopédie de l'Utopié, des voyages extraordinaires et de la Science Fiction.* Lausanne, 1972
39. **Joseph Weizenbaum.** *Computer Power and Human Reason – from Judgment to Calculation.* San Francisco, 1976
40. **Norbert Wiener.** *The Human Use of Human Beings.* New York, 1970
41. **Bernard Wolfe.** *Limbo '90.* London, 1953
42. **John F. Young.** *Robotics.* London, 1973
43. **Yevgeny Zamyatin.** *We.* London, 1972

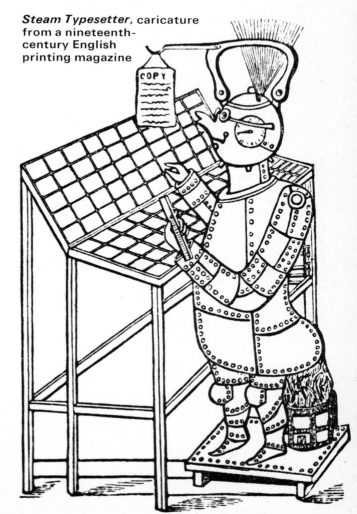

Steam Typesetter, caricature from a nineteenth-century English printing magazine

Index